CSR, Sustainability, Ethics & Governance

Series editors
Samuel O. Idowu, London Metropolitan University, London, United Kingdom
René Schmidpeter, Cologne Business School, Germany

More information about this series at http://www.springer.com/series/11565

Hualiang Lu • René Schmidpeter •
Nicholas Capaldi • Liangrong Zu
Editors

Building New Bridges Between Business and Society

Recent Research and New Cases in CSR, Sustainability, Ethics and Governance

Springer

Editors
Hualiang Lu
Sino-German Research Institute for Sustainable Development
Nanjing University of Finance and Economics
Nanjing, Jiangsu
China

René Schmidpeter
Cologne Business School
Cologne, Germany

Nicholas Capaldi
College of Business
Loyola University New Orleans
New Orleans, Louisiana
USA

Liangrong Zu
International Training Centre of the IL
International Labor Organization
Turin, Italy

ISSN 2196-7075 ISSN 2196-7083 (electronic)
CSR, Sustainability, Ethics & Governance
ISBN 978-3-319-63560-6 ISBN 978-3-319-63561-3 (eBook)
DOI 10.1007/978-3-319-63561-3

Library of Congress Control Number: 2017956136

Printed on acid-free paper

This Springer imprint is published by Springer Nature
The registered company is Springer International Publishing AG
The registered company address is: Gewerbestrasse 11, 6330 Cham, Switzerland

Preface

Businesses play an important role in society and impact communities, regions, as well as individuals and the environment. The processes and consequences of globalization and the economic crisis have repositioned business approaches and growth toward more inclusive, competitive, sustainable, resource- and energy-efficient, and environment-friendly management. Business management models should adequately address the key objectives of inclusive and sustainable management, integrating economic, social, and environmental elements and considering the inter-linkages and collective impact in all three dimensions. There is an urgent need to formulate sustainable management frameworks addressing interactions and dependencies as well as trade-offs in various focus areas. The challenge that these adjustments present calls for the alignment of corporate social responsibility (CSR) with business strategy to create a truly sustainable approach.

In a global economy, no business or form of commerce is an island unto itself. Rapid and dynamic changes in technology, markets, political and legal institutions, and indeed cultures have all created new challenges. In order to overcome the myopia of limited perspectives, a new organization has been created: Global Corporate Governance/Corporate Social Responsibility. The intention is to bring together scholars from many fields along with business, academic, cultural, religious, and political leaders to form a global alliance dedicated to rethinking and integrating value issues into management practice, education, and development.

Toward this end, series conferences, seminars, and workshops are organized worldwide. The first annual conference in this regard was organized in London, UK, in 2014 by launching the newly born CGC/CSR organization. The second year's conference was held in Nanjing, China, in 2015. The next two annual conferences will be held in Cologne, Germany, and Perth, Australia, respectively. For each conference, the best papers will be selected and published in a Springer book.

The conference, seminars, and workshops will showcase current CSR models and practices, as well as the next generation of issues that business leaders and society will face. It provides a platform for the sharing of expectations, aspirations, and responsibilities. It highlights state-of-the-art topics and how issues are being addressed around the world. In addition, we aim to develop new frameworks, tools, and techniques essential for the integration of socially responsible management in business operations, in an effort to achieve sustainability at all levels of business management.

Nanjing, China — Hualiang Lu
Cologne, Germany — René Schmidpeter
New Orleans, LA, USA — Nicholas Capaldi
Turin, Italy — Liangrong Zu

The Aim of This Book

The aim of this book is to provide a comprehensive understanding of the linkages between business and society by covering key issues in various themes of corporate social responsibility, sustainability, ethics, and governance, thanks to the different visions and perspectives offered by the authors from different countries and having different expertise.

More specifically, this book aims to deepen the understanding of the up-to-date knowledge and theories in school of corporate social responsibility, sustainability, ethics, and governance. It also wishes to provide practical solutions when businessmen and practitioners apply CSR and sustainability with business strategies and management. Furthermore, this book aims to be an international think tank thanks to Global Corporate Governance/Corporate Social Responsibility Institute bringing together scholars from across the globe.

Introduction

Entrepreneurship's Relationship to CSR (Prof. Dr. Stephen R.C. Hicks)
This chapter rethinks the start of business ethics. The author agrees that the Corporate Social Responsibility model of business ethics has been a leading paradigm. But the author notices that practitioners usually take large firms as representative of business and address their ethical issues; this, he believes, leads to overgeneralizing. But most people do not work in mid to large corporations; rather, they are sole proprietors, in a partnership, in a family firm, or in an entrepreneurial venture. Also, every large corporation began as an entrepreneurial venture. Therefore, the author argues that business ethics should begin where business begins. In other words, business ethics begins with entrepreneurship. The author first situates ethics in an entrepreneurial context to identify the core values, virtues, and vices of business. Then he addresses how those ethical issues scale as the business succeeds or fails at growing into large corporation.

In the Pursuit of Building the Foundation for Sustainability (Nayan Mitra)
The author recognized that a corporation exists in time and space which will take toward sustainability when managed responsibly. She agrees that sustainability comprises three broad panaceas, namely, social, environmental, and economic responsibility, also known as the three pillars (people, planet, and profit) of sustainability. But she argues that it is not enough to only think of the corporation (or the focal company) when studying sustainability. Thus, this chapter extends from existing study focusing on corporation to supply chain and studies the dynamics of sustainability by taking examples from India.

Risky Business? On the Interplay Between Social, Actuarial, and Political Risks and Licenses (Martin Brueckner, Sara Bice, and Christof Pforr)
This chapter extends from previous well-established social and actuarial licenses to introduce the third license, political license. The authors agree that corporate social responsibility (CSR) and social license to operate (SLO) have become cornerstones of many companies' risk management strategies. But the authors argue that there are gaps in understanding the concepts of CSR and SLO not only of their commonalities

and differences but also of their politicality and conflicted nature. This triad of licences is placed within a dynamic risk framework that helps progress the CSR and SLO discourses from typical organizational risk management approaches and provides a more holistic conceptualization of the field of licenses to be navigated and negotiated by all SLO/CSR stakeholders. This approach can serve as a foundation for critical research in the CSR and SLO space, enabling the analysis of, and discussion on, the meaning, intention, and probable implications of the various, at times competing, types of licences and explicating some of the conceptual weaknesses that have long plagued both scholarly fields.

Sustainable Logistics: A Framework for Green Logistics and City Logistics (Prof. Dr. Carsten Deckert)
The chapter proposes a framework for sustainable logistics based on the research for German book project. The author is aware of both positive and negative effects of logistics. With the further economic development of urbanization, there is a need to tackle the challenges to logistics by applying green logistics and city logistics. The author proposes the sustainable logistics covering three logistical functions: transportation, warehouse management, and packaging. In order to achieve sustainable objective, all logistics functions should be organized in a sustainable manner. And trade-offs between the functions of transportation, warehousing, and packaging have to be put into focus in the future.

Sustainable Assortment Policy: Possibilities of Differentiation and Profiling for the Food Sector (Prof. Dr. Christoph Willers and Victoria Aydin)
This chapter introduces the concept of sustainable assortment policy for food sector to differentiation and profiling. The authors argue that anybody who includes the "sustainability" in its corporate or product brand is able to gain a real competitive advantage. The challenge on commercial side therefore consists in putting food on the market whose social-ecological production is clearly proved and which consumers can trust. The authors suggest that trading ventures should consider the indicated trends in the complex theme of sustainable products to take the opportunity and at the same time reduce potential risks. As a result, individual measures are no longer sufficient to communicate the competences concerning sustainability—rather, innovative solutions as well as credible and suitable overall concepts are required for food sector.

The Importance of Gold in the Financial Report (Prof. Dr. Rute Arbeu and Carlos Pinho)
This chapter focuses on the identification, prevention, and mitigation of the adverse impacts of the gold as an asset through the financial report. The authors aim to contribute to raising awareness on the integration of financial reporting and demonstrate how accounting issues are a concrete dilemma of understanding the gold as an asset. The authors declare that there is a need of more transparency, and the increased awareness of the communication process of the real state of the economy and the budget state is an attitude of willingness to learn from past mistakes and then cooperate with society to improve and promote the economic growth.

Sustainable Hospitality Management: Challenges and Opportunities for Small Island Destinations: Lessons from British Virgin Islands (Bonnie Lewtas)
This chapter determines sustainability challenges for small island destination hotels and how obstacles can be overcome or avoided through innovative and integrative management techniques. The author develops a broad understanding of sustainable hospitality management. The author argues that key areas for small island destination hotels, for example, building and construction, purchasing of building materials, sustainable transportation, guest activities, and the social dimension, are not currently considered. The sustainable management of small island destination hotels is diverse, sustainability management is at the core of long-term business success, and every initiative undertaken directly or indirectly supports the long-term viability of hotels.

Accounting for Sustainability: The Case Study of Petrobras (Prof. Dr. Rute Abreu, Fátima David, Liliane Segura, Henrique Formigoni, and Flávio Mantovani)
This chapter demonstrates the adoption of the appropriate accounting framework supported on the International Accounting Standards and the influence of International Auditing Standards on the promotion of a corporate governance strategy approved by companies to increase the transparency of the information. The authors focus on the professional skepticism in relation to the fairness of assertion on the financial statements and the management responsibility for global interest in corporate responsibility and sustainability. The authors argue that the accounting and financial reports will be the weapons to combat malfunctioning of the financial markets.

Mineral Supply Chain Transparency: Soft and Hard Laws on Supply Chains Due Diligence and the Rise of Public–Private Partnerships (Dr. Fabiana Di Lorenzo)
This chapter looks at soft (voluntary) and hard (mandatory) laws on due diligence and supply chain transparency in the mineral sector and the rise of multistakeholder initiatives or public–private partnerships (PPPs) as a means to promote legal advancement and compliance. Focusing on responsible mineral supply chains, it argues that since the Human Rights Council endorsed the UN Guiding Principles on Business and Human Rights in 2011, the path to regulate businesses' behavior turned toward mandatory legislation while keeping voluntary forms of regulation. The greater call for due diligence in mineral supply chains has also been accompanied by a greater interest in PPPs to promote companies' compliance with human rights, environmental standards, and laws. The chapter concludes that PPPs are becoming agents for legal advancement and a prominent feature in the respect and protection of human rights along supply chains, as agents of human rights advancement, legal enforcement, and legal efficiency. It also emphasizes that in order to succeed, PPPs need the full engagement of governments of mineral producing countries whose role should not be limited to the mere provision of a regulatory framework but include the creation of a supportive environment for the full implementation of due diligence.

CSR, Innovation, and Human Resource Management: The Renaissance of Olivetti's Humanistic Management in Loccioni Group, Italy (Dr. Del Baldo Mara)
This chapter addresses the theme of corporate social responsibility policies concerning labor and employees. The author proposes an anthropologically centered analysis of business and leadership models that are geared toward a multidimensional development. The case study of Loccioni Group offers an example of best stakeholders' and employees' management practices, which coevolves with the environment, improving the company's competitiveness and the socioeconomic conditions of the local context. The author concluded that CSR is part of the DNA and widespread throughout the entire organization. Loccioni Group helps develop reflections on the importance of embracing the cultural and anthropological roots of CSR and reinvent the Olivetti's model of holistic development while conceiving the business as a tool for promoting social, economic, moral, and environmental well-being.

Does Foreign Ownership Enhance the Corporate Social Performance of Japanese Firms? (Prof. Dr. Magumi Suto and Prof. Dr. Hitoshi Takehara)
This chapter reports the research of corporate social performance in Japan. With the rapid changing ownership structure and globalization, the authors investigated the influence of foreign ownership on the corporate social performance. The authors declare that foreign investors play an important role in shifting Japanese corporate governance from the traditional insider-oriented structure to a structure that is characterized by greater openness and transparency to survive and success in global competition. They concluded that foreign investors make Japanese firms to improve CSP by motivating firms to reconsider trustworthiness of their business in global society and markets and change their corporate social responsibilities.

Study on the Impacting Factors of Triple Performance for Farmers' Professional Cooperatives: A Case Study of Jiangsu Province (Prof. Dr. Hualiang Lu, Dongqin Wang, and Mengya Wang)
This chapter studies economic, social, and environmental performance of cooperatives in China. The authors develop an integrative research framework covering internal and external environment, production operation, governance structure, and performance. Applying the structure equation modeling, the authors discovered that external industrial environment and good cooperation relationship between cooperatives and the local leading enterprises will significantly affect economic and social performance of cooperatives in China via the effect of governance structure. Furthermore, the level of services provided by the cooperatives and the level of unified management will contribute to the environmental performance of cooperatives. The chapter ends with several suggestions for sustainable development of cooperatives in China.

Integrative Model in Mitigating the Impact of International Labor Migration on Family Left Behind: Case Study in Indramayu, Indonesia (Lina Widyastuti)
The author provides an integrative model between government, nongovernment organization, state-owned bank, and communities in addressing the impact of international labor migration on the family left behind. She argues that the communities should take an important position in this model because they are able to actively involve in the implementation of the projects in both government and private sectors. Besides, government, nongovernment organization, the companies, and bank should take their own responsibilities.

CSR in the Context of Transition Economy: An Evaluation of Enterprises CSR Practices in China (Bing Zhu and Prof. Dr. Andre Habisch)
This chapter reports the CSR practice in China and how it has grown in the past years. The author argues that the ownership takes a major role in applying corporate social responsibility practice. He affirms that big companies (e.g., central and local state-owned enterprises) are the main force of CSR practice in China. Moreover, he explains that corporate social responsibility has received recognition and application in various industries more than ever before, and CSR practice becomes the mainstream of sustainable development in China.

Women Symbolism in Marketing: Are the Human Rights Legit? (Debarati Das Gupta and Aruna Das Gupta)
This chapter focuses on the women's rights in Indian. They argue that the Indian Constitution prohibits discrimination on the basis of sex, but the position of women in India remains unequal. The authors start from several Ads to prove that the basic human rights of women in India were violated through unnecessary marketing objectifications. They argue that a lot of sectors (media, society, corporate sector) have to conceive together a notion of responsibility as it affects in the procurement, settlement, and acceptance of the image they sell in advertisements. They declare that parties in question must work with Public Policy–Private Sector partnership and employ Fundamental Rights, especially Human Rights. They concluded that a mutual partnership between the Public Policy and Private sector shall strengthen India's fight for Women's Rights and Gender Equality.

Contents

Part I
CSR Origins

Entrepreneurship's Relationship to CSR

Stephen R.C. Hicks

1 Three Venture Capitalists on Virtue and Business

Testifying before a U.S. Congress House committee, the financier J. P. Morgan (1837–1913) replied to a question asking whether he loaned out money based on the recipient's assets. Morgan replied, "No, sir, the first thing is character." And, he continued, if someone asked for funding but Morgan felt he couldn't trust him, he wouldn't make the loan even if he had "all the bonds in Christendom."

The venture capitalist Georges Doriot (1899–1987) was described by biographer Jeffrey Young this way: "Doriot spends most of his time talking to people who bring him prospective investments. He says he has considered no less than 5000 of them since 1946. He is considered by friends and critics alike as a brilliant judge of character. But he has to be, he explains. 'When someone comes in with an idea that's never been tried, the only way you can judge is by the kind of man you're dealing with.'"

Entrepreneur and venture capitalist Kevin O'Connor (1961–) identifies what he looks for when evaluating entrepreneurial pitches: "I look at three big things: Have they found a big problem in a big market? Have they solved the problem in what I believe to be the most effective way? And are they able to pull it off—are they smart, are they aggressive, are they honest, and hard-working?" (O'Connor 2009).

What the three entrepreneurs' statements have in common is their making *character* fundamental to business.

S.R.C. Hicks, Ph.D. (✉)
Rockford University, 5050 East State Street, Rockford, IL 61108, USA
e-mail: SHicks@Rockford.edu

H. Lu et al. (eds.), *Building New Bridges Between Business and Society*, CSR, Sustainability, Ethics & Governance, DOI 10.1007/978-3-319-63561-3_1

2 Entrepreneurship and Ethics

Ethics is about values and virtues. What goals in life are worth pursing—are health, wealth, love, beauty, creativity, and so on, *values*? And what character traits are necessary to achieve those good things—are honesty, courage, perseverance, fairness, and so on, *virtues*?

In the past generation, while more attention is being given to entrepreneurship as an economic phenomenon,[1] entrepreneurship has received much less attention as an ethical phenomenon. Yet entrepreneurship is a value-laden enterprise. As a business activity, entrepreneurship demands resourcefulness and resilience, risk tolerance and courage, and it can be a vehicle for achieving important financial and other personal goals.

In this essay, I sketch an entrepreneurial ethics, explore its implications for the foundations of business ethics, and contrast it to a currently dominant paradigm for business ethics, i.e., Corporate Social Responsibility.

3 The Values of Entrepreneurship

What does it mean to see entrepreneurship as an ethical phenomenon?

Let's approach answering the question by first asking why we work. Some people work because they must; others work because they want to. Unless one is born into wealth or wins the lottery, making a living is foundational to life. Hence most of us must work to sustain our lives at least to a minimum level. But beyond a bare making a living, work is be a means of achieving many goals—financial security, creative expression, sociability, and even adventure. Many of those who do not have to work often find their work to be a vehicle for pursing those goals.

One basic work choice is to go to work for an existing business, i.e., to become an employee, or to go into business for oneself, i.e., to become an entrepreneur. So let us ask, secondly: Why do some choose entrepreneurship over working for others? Their reasons can be negative—there may be no employment opportunities available so entrepreneurship is the best available option. Or their reasons can be positive—they believe that as entrepreneurs they can achieve greater wealth, autonomy, and self-expression.

Wealth Entrepreneurs typically have ownership positions in their businesses and thus a chance at greater earnings than salaried or hourly employees. That wealth in turns means a chance at greater financial security and all the good things that money can buy.

---

[1]Baumol (2010) notes: "Entrepreneurs are widely recognized for the vital contributions they make to economic growth and general welfare, yet until fairly recently entrepreneurship was not considered worthy of serious economic study."

Autonomy Entrepreneurs are deciders and initiators. Consequently, they have greater power over their thoughts and actions and more flexibility in the use of the time than most employees. Expressed negatively, autonomy means not having to take orders from bosses, being more easily able to avoid bureaucracy, office politics, and so on.

Self-expression Entrepreneurs build businesses from the ground up, typically generating the idea and making it happen. As a result, entrepreneurs typically experience stronger psychic ownership of their businesses. Their business is *My idea done my way*. Put negatively, entrepreneurs do not experience their work life as being a cog in someone else's machine.

Entrepreneurship, accordingly, is a vehicle for the pursuit of several important physical and psychological human values: wealth, autonomy, and self-expression. Of course, as an employee one can realize those values, yet entrepreneurship makes those values central and more likely to be realized.

Further: entrepreneurs trade with others when purchasing raw materials, when selling to customers, and when they hire employees to help grow the business. Entrepreneurs create new networks of voluntary traders, and they enter into and maintain already-existence networks. Trading networks are also a great human value, and they are based upon mutual commitments to deal with each other on the basis of voluntary exchange, rather than, say on the basis of treating the other as an object of charity or plunder.

Accordingly, entrepreneurship is based upon several value commitments: the importance of autonomy, self-expression, voluntary trade, and the creation and enjoyment of wealth.

4 The Virtues of Entrepreneurship[2]

If entrepreneurship is a value-oriented enterprise, then what virtues make that enterprise successful? That is to say, if one is to achieve practical business success, what character traits must one strive to embody and make habitual?

Let us consider a typical entrepreneurial process, and note the italicized concepts in the following description. Entrepreneurship begins when someone has a *new business idea*.[3] The entrepreneur is *ambitious* and exhibits *guts* in taking the *initiative* in developing the idea into a new enterprise.[4] Typically through *experimentation* and *perseverance*, the entrepreneur *produces* something of value. The entrepreneur then becomes a *leader* by convincing customers of the new product's value and by teaching new employees how to make it. The entrepreneur and

[2]This section builds upon Hicks (2009).

[3]Kirzner (1973) makes this aspect of the entrepreneurial process the defining aspect.

[4]Knight (1921) makes this aspect of the entrepreneurial process the defining aspect.

customers and the entrepreneur and the employees *trade to win-win* results. The entrepreneur has achieved a measure of *success* and then is able to *enjoy* the fruits of his or her achievement.

What does this sketch of the entrepreneurial process have to do with virtue? Virtues are action-guiding character traits that aim at good results.

So if we cash out the above italicized entrepreneurial-process traits in terms of virtues—i.e., in terms of character traits and commitments that enable and constitute good action—then we make the following connections:

The entrepreneur's generating *new business ideas* connects to the virtue of *rationality*. Rationality is the commitment to the exercise of one's capacity for reason. The entrepreneur's coming up with a business idea, evaluating it, and planning to make it real require the exercise of rationality.

The entrepreneur's *ambitious* drive for success connects to the virtue of *pride*. Pride can be based upon past accomplishments, but it can also be future-oriented—wanting to be the best one can be and not settling for less. The entrepreneur's ambition is his or her taking pride in the business part of his or her life.

Entrepreneurial *initiative* connects to the virtue of *integrity*. Integrity is a commitment to acting on the basis of what one believes to be true and good. If the entrepreneur believes a business idea to be good, then the activity of making the ideal a reality *integrates* thought and action.

The risk of failure is a feature of entrepreneurship, and fear is a natural response to the possibility of failure—of losing money, losing self-esteem, and losing respect from others. So the *gutsiness* that entrepreneurial action involves connects to the virtue of *courage*. Courage is acting to achieve what one thinks is good even when one is aware of the possibility of failure.

The entrepreneur's working through *experimental* process of product development connects to the virtue of *objectivity*. To be objective means judging based on one's awareness of the facts, being open to new data (including unwanted negative data). An element of objectivity is the virtue of intellectual *honesty*, that is, a commitment to recognizing the facts of the matter for what they are.

The entrepreneur's *perseverance* through difficulties, disapproval, and self-doubts connects with the virtue of *independence*. Independence is the virtue of trusting one's own judgment and acting on the basis of one's best judgment despite frustrations, distractions, or the negative opinions of others.

The entrepreneur's *productivity*, i.e., his or her sticking with it until the job is done, connects to the virtue of *productiveness*. Productiveness is a commitment to creating value, to being self-responsible for bringing into existence that which one needs or wants.

The network of traits that make for effective *leadership* are complex and variable. Yet through whatever individual capacities and personality traits, entrepreneurs must demonstrate to others the value of the new business's products or services, convince customers and investors to commit funds, and teach employees how to perform the business's functions. Entrepreneurship necessarily involves a commitment to *leadership*.

Table 1 Entrepreneurial traits and moral virtues

Entrepreneurial traits	*Moral virtues*
New ideas	Rationality
Ambition	Pride
Guts	Courage
Initiative	Integrity
Perseverance	Independence
Experimentalism	Objectivity (including honesty)
Productivity	Productiveness
Leadership	Leadership
Win-win trade	Justice
Entrepreneurial consequences	*Moral values*
Experiencing success: wealth, autonomy, self-expression	Self-esteem, happiness, flourishing

The entrepreneur's *trading to win-win results* with customers and employees connects to *justice*. The virtue of justice entails a commitment to evaluating and interacting with others according to merit. In a business context of voluntary deals, justice means that each party judges the merits of trade independently and agrees voluntarily to the terms of the trade, and follows through accordingly.

And, finally, the entrepreneur's *achieving and enjoying success*, both the physical and psychological rewards that business achievement can bring, connect to the general *moral values of flourishing, happiness, and fulfillment*. As the entrepreneur's business life is a component of his or her overall life, the entrepreneur's engaging in the actions that lead to flourishing in business is a component of an overall flourishing life. In Aristotelian terms, the entrepreneur's actions both constitute and lead to a life that is fully realized.

Summarizing all of the above in a table, we get the following (Table 1).

5 Entrepreneurial Ethics as a Business Ethics Code

The virtues constituting the table's right column embody an entrepreneurial code for business ethics. That set of virtues describes entrepreneurial activity abstractly. Conversely, the success traits of entrepreneurs in the table's left column are particulars of a general set of virtues that can be applied in all walks of life.[5]

Relevant questions, then, for the business ethics of entrepreneurship are:

(a) What are the *values* of entrepreneurship, upon which all successful business is based?
(b) What are the character *virtues* of entrepreneurship that enable successful business?

[5]This list of virtues integrates those found in Aristotle (1984) and Rand (1964).

(c) How do we *teach and inspire* those values and virtues in students and business professionals?

Now for some implications.

One implication is that approaching business ethics via entrepreneurship connects business to ethics positively and organically. The virtues and values embedded in the practice of entrepreneurship sets a foundation for a business ethic based on the assumption that successful business practice has within it the resources to develop an ethic. That approach contrasts to the assumption often made that ethics is alien to business and must be grafted on or imposed from without.

Another implication of an entrepreneurial ethic is that business ethics should focus first on creativity, productivity, and trade. That is, it should not take those elements take them for granted or as amoral givens. If the basis of business is creative productivity and trade, then the basis focus of business ethics should be upon that which enables individuals to be or become creative producers and traders.

A third set of implications emerge when we contrast an entrepreneurial business ethics to the dominant model of business ethics for the past half-century, i.e., "corporate social responsibility" (CSR).

6 Entrepreneurial Ethics Contrasted to Corporate Social Responsibility

Corporate Social Responsibility's three constituent words indicate its three framing assumptions as a model of business ethics:

The first tells us that *corporations* are our model of business practice to analyze and prescribe to.

The second word tells us that the *social* is our focus.

In the literature, *responsibility* is usually cashed out in terms of *avoiding harm* and *distribution to others*. Or it is interpreted conjunctively with the second word to mean *social* responsibility.

On Corporations as the Standard Model The vast majority of examples in business ethics focus on large, well-known corporations—Walmart, Microsoft, Enron, McDonald's, and so on—and much of the business governance literature focuses upon corporate governance, with its hierarchical structure.

Yet the corporation is not the only business type or even the most plentiful type. Firms can be organized as sole proprietorships, partnerships, or corporations, and US Census Bureau data indicate that large corporations are a fraction of overall business activity: "About three quarters of all U.S. business firms have no payroll. Most are self-employed persons operating unincorporated businesses" (U.S. Census Bureau 2007). Further, in 2007 there were 27,757,676 firms in the U.S., but the number of firms with more than 100 employees was approximately 126,000, which is about one-half of one percent.

What these numbers suggest is that the business ethics literature's focus on mature, large corporation obscures the reality of the business environment that most people experience. Most businesses are small businesses and closer to their entrepreneurial roots.

Further, every corporation begins as an entrepreneurial venture. Many such ventures perish, a few go on successfully, and only a very few become large. But there is a value to understanding businesses causally. The basic principles of business—buying and selling, hiring, firing, and quitting—including its moral principles, are in place from the beginning. So a focus on entrepreneurship enables us to articulate those moral principles at the foundation of business activity and then to see how they develop as the business grows in size and complexity.

On the Social Focus CSR typically assumes, as in the following example from the Committee on Economic Development, that business exists to serve social purposes, often top-down social purposes: "business functions by public consent and its basic purpose is to serve constructively the needs of society—to the satisfaction of society" (Carroll 1999). Such formulations do not mention individuals and assume that the social has priority over the individual.

Yet it is not clear that this is the reality or importance of business. The basic business transaction is that between a buyer and a seller: two individuals come together, each bringing value to the transaction, and they go their separate ways after to enjoy the results of the transaction. Before buyers and sellers can trade, each has to engage in productive work, and productive work is basically individual. An individual wakes up in the morning, decides to get out of bed and go to work. He or she makes dozens of decisions and performs hundreds of particular actions in the course of a day. At work, those actions add up to productivity, the results of which enable the person to become a buyer or a seller.

Entrepreneurial ethics highlights this individuality of business: entrepreneurial activity is primarily individual. It starts with an individual's idea, an individual's productive efforts, and then develops into a social network of value-adding individuals producing and trading with each other. What we call firms, markets, and other relevant social groupings are associations of individuals adding value together. Such social groupings are formed bottom-up and remain in existence as long as they serve mutually the interests of the individuals involved.

So while social interaction is an important part of business ethics, it is consequent. A business ethic based on entrepreneurialism makes issues of individuality primary and issues of sociability secondary.

On Responsibility as Avoid-Harm and Distribution In CSR, the broad concept of responsibility is typically given two sub-meanings. One meaning is avoiding harm to others—e.g., not engaging in fraud, not damaging others' property, and so on. The other is to engage in charitable or redistributionist activities—for example:

"Traditionally in the United States, CSR has been defined much more in terms of a philanthropic model. Companies make profits, unhindered except by fulfilling their duty to pay taxes. Then they donate a certain share of the profits to charitable causes. It is seen as tainting the act for the company to receive any benefit from the giving" (Baker 2011).

While there certainly are places for anti-harm and philanthropic principles in ethics, from an entrepreneurial perspective this concept of responsibility is narrow and secondary. CSR typically makes no explicit mention and certainly does not emphasize the priority of productivity as a moral responsibility.

By contrast, the focus of entrepreneurs first and foremost is upon production, not philanthropy; and entrepreneurs are focused upon creating value, not avoiding harm. *How can I create value?* is the operative question. Or in other words: *How can I be more productive?* or *How can I make money?* (with an emphasis upon the *make*). Creating, producing, and making are primary.

One reason for this is that causally production comes before distribution. Before we can ask the distribution question of *Who gets what?*, the *what* has to be brought into existence. That is the basic responsibility. Before one can make a distribution claim on the value a business has created, one must have productively contributed to the creation of that value. The issue of distributive justice falls out of the productivity: A fair determination of who gets how much depends upon each person's productive contribution.

While entrepreneurial ethics makes our responsibilities to be creative producers primary, CSR's emphasis on distributive responsibilities, especially charitable distributions overlooks the morally and causally prior productive responsibilities.

The same point about moral priorities applies to the avoidance of harm. Part of successful productivity is not harming the legitimate interests of others while achieving one's own goals, but the focus is on the positive creation not the non-harm. By analogy, if one's goal is to travel from New York to Los Angeles, one's focus is upon successful transportation; one's primary focus is not upon not harming Nebraskans along the way.

To summarize in abstract terms, the avoid-harm principle says that we should *not* engage in win-lose interactions with others, and the redistribution principle says that we *should* engage in lose-win interactions. What is missing is the great moral import of making possible win-win interactions.

Entrepreneurial ethics makes first our responsibilities as individuals to be productive traders. The dealings of productive traders are neither harmful (win-lose) nor charitable (lose-win). Instead, they are win-win.

The morality of productiveness is prior to the morality of distribution. So business ethics should be focusing first and predominantly on self-responsible productiveness and the social conditions that foster it.

7 Conclusion

While entrepreneurship is beginning to receive attention in the business ethics literature, it has the potential for refocusing our attention to business fundamentals. The questions, in the contrasts between traditional CSR and entrepreneurial ethics, are:

Should we focus first upon *start-ups* and innovative firms, or upon *mature corporations*?
Should we focus upon *individuals* first, or upon the *social*?
Should we focus upon *production and trade* first, or upon *harm-avoidance and charity*?

What entrepreneurial ethics suggests is: Do not start business ethics with corporations. Do not start with the social. And do not start with philanthropic accounts of responsibility. Instead, start by making morally foundational those productive individuals who engage in win-win trade, and build from there to the increasingly complex social structures the business world creates.

References

Aristotle. (1984). Nicomachean ethics. In J. Barnes (Ed.), *The complete works of Aristotle*. Princeton: Princeton University Press.

Baker, M. (2011). Corporate social responsibility—What does it mean? Accessed May 8, 2010, from http://www.mallenbaker.net/csr/definition.php

Baumol, W. J. (2010). *The microtheory of innovative entrepreneurship*. Princeton: Princeton University Press.

Carroll, A. B. (1999). Corporate social responsibility: Evolution of a definitional construct. *Business Ethics Quarterly, 38*(3), 268–295.

Hicks, S. (2009). What business ethics can learn from entrepreneurship. *Journal of Private Enterprise, 24*(2), 49–57.

Kirzner, I. (1973). *Competition and entrepreneurship*. Chicago: University of Chicago Press.

Knight, F. (1921). *Risk, uncertainty, and profit*. Boston: Houghton Mifflin.

O'Connor, K. (2009). Interview on "entrepreneurship and venture capital." *Kaizen, 6*. Accessed May 3, 2010, from http://www.ethicsandentrepreneurship.org/20090429/interview-with-kevin-oconnor/

Rand, A. (1964). The objectivist ethics. In The virtue of selfishness. New York: New American Library.

United States Census Bureau. (2007). Accessed May 18, 2010, from http://www.census.gov/econ/smallbus.html

In the Pursuit of Building the Foundation for Sustainability

Nayan Mitra

1 Introduction

A company (also called, Corporation, Organization, business, firm in this paper) needs to go beyond mere existence, survival and growth to perpetuate. It is this concept of perpetuation or going concern, that has often made me think. How is this possible?

A Corporation is a social entity. It has its existence in time and space; it has its upwards and downwards dynamics that binds it together to form a whole that should be able to survive and perpetuate in a very competitive environment. Therefore, if we talk about the focal Company alone, the discussion will be incomplete. If a Company complies to sustainability norms, it is wonderful; but if the company's supply chain flouts the sustainability norms and the company uses its products and services, is it still okay? Will there not be a lacuna then in their 'responsibility' to the society or in its issue of perpetuity?

2 Objectives

This paper, therefore seeks to delve deeper into:

1. The dynamics of sustainability in its upward integration, that is, the supply chain,
2. Point out its relevance, and
3. Ways and means of management;

with references and examples drawn from the Indian context.

N. Mitra (✉)
Developmental Consultant, Kolkata, West Bengal, India
e-mail: mitra.nayan@gmail.com

H. Lu et al. (eds.), *Building New Bridges Between Business and Society*, CSR, Sustainability, Ethics & Governance, DOI 10.1007/978-3-319-63561-3_2

Now, the question is: what comprises of the supply chain? Christopher (2012) defines supply chain as "*the network of organisations that are involved through upstream and downstream linkages, in the different processes and activities that produce value in the form of products and services in the hands of the ultimate consumer.*"

Thus, supply chain comprises of a whole ecosystem from the minutest supplier to the final customer and, any non-compliance or irresponsibility in this milieu will ultimately affect the focal company's sustainability, at the end. This supply chain, therefore, needs to be managed to ensure responsible governance, traceability and hence, accountability.

Supply chain Management, as defined by the **Association for Operations Management (APICS)** is "*the design, planning, execution, control and monitoring of supply chain activities with the objectives of creating net value, building a competitive infrastructure, leveraging worldwide logistics, synchronizing supply with demand, and measuring performance globally*" (Wisner et al. 2011).

3 Conceptual Framework

The conceptual framework of this paper is as follows (refer Exhibit 1):

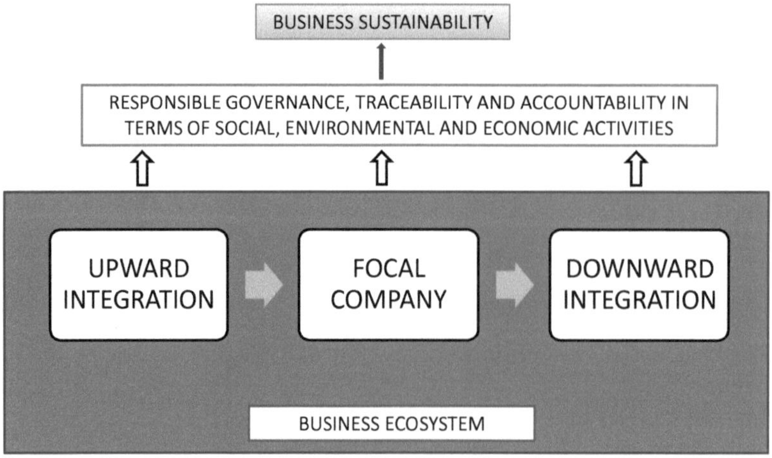

Exhibit 1 The foundation for business sustainability

4 Literature Review

Modern supply chains, in this age of globalisation and dynamic trade practices, are, in many cases, multi-tiered networks that involves various types of suppliers across various demographical, socio-political and geographic spread (Giblin 2013).

Therefore, in order to have an equilibrium between the focal company and its business ecosystem, there needs to be a uniformity in some of their key corporate policies and interactions that bind them together in linearity.

Here comes the debate, where, academicians and particularly practitioners have often questioned the scopes, limitations and extent of the focal companies' role in ordering its suppliers and even their suppliers, to comply (Gilmore 2015).

Let us critically analyse and discuss the various sustainability concerns faced by the modern corporations.

4.1 Social Concerns

Vogel (2005) have noted that while prominently featuring information about their codes of conduct, some companies often ignore reporting the extent of compliance with them; while some others describe their own practices, but ignore those of their supply chain. Let us follow some of the Indian examples given below (Refer Table 1).

Given the above examples, in Table 1, who suffers? If the supply chain does not adhere to sustainable practices, is it not the focal company, whose brand is at stake? Is it not, always the company's long-term sustainability that comes under question?

Infact, ***Anand Mahindra (2010)****, the Vice Chairman and Managing Director, Mahindra and Mahindra Limited,* an Indian industrial group *feels*, "The purchasing power of a corporation can become a unique driver for bringing about positive change in society. Companies must use this power to achieve a purpose and make their supply chain a vehicle for inclusive growth" (UNGC, 2010). The Company initiated a programme to enable knowledge and best practice sharing with suppliers in order to address environmental, health and safety impacts throughout the company's supply chain (UNGC 2010).

4.2 Environmental Concerns

Among environmental concerns, the second panacea of sustainability, climate change is one of the major global concerns. Most climate scientists agree the main cause of the current global warming trend is human expansion of the "greenhouse effect," (GHG) caused by a build-up of greenhouse gases, primarily carbon dioxide,

Table 1 Lack of supply chain sustainability in social concern issues—Indian examples

Year	Company	Issue	Steps taken	Outcome
June, 2008 (Daily Mail Reporter 2008)	**Primark**, UK and Ireland's budget chain known for its low-cost, high fashion clothing range supply chain	– Children as young as 11 were working in squalid conditions, sewing tiny beads and sequins onto cheap t-shirts by candle-light in the store's Indian sweatshops at a maximum daily wage of INR 40 (much less than one US $ per day) (revealed through an undercover investigation done by the BBC Panorama team for 6 months)	The Company claimed shock and discontinued with three suppliers in India for passing work to unapproved sub-contractors using child labour	– Bad publicity, plunging profit, came to be known as the 'sweat-shop brand'
2007 (McDougall 2007)	**GAP**, the international garments and accessories chain	– Children working in filthy conditions in the Shahpur Jat area of Delhi, India (discovered by The British newspaper, The Observer)	The company pledged to convene a meeting of its Indian suppliers as well as withdrawing tens of thousands of the embroidered girl's blouses from the market, before they reach the stores	– Bad publicity, financial and non-financial loss to Company

methane, and nitrous oxide that traps heat in the atmosphere, radiating from Earth toward space (Global Climate Change 2015).

The **2014 report from the Carbon Disclosure Project (CDP)**, an independent global system for companies to measure, disclose, manage and share climate change and water information, which tracks the GHG emissions and the carbon footprint of not only their own manufacturing activities, but also their transportation, distribution and procurement activities, and monitors the related activities of their extended supply chains (Blanchard 2012), has, for the first time, put five Indian companies, namely **Wipro, Essar Oil, Tech Mahindra, Larsen and Toubro and Tata Consultancy Services** on its Global A-list (Clough 2015).

This is indeed a great news. The consciousness has begun.

Our collective greenhouse gas emissions will dictate whether or not we risk tipping the world towards dangerous climate change, that will change our dynamics of business, among a lot many other things (Supply Chain Report 2015). Thus, the foundation of the business, in terms of its supply chain, needs to be strong with respect to their own sustainability issues to make the focal company stronger and sustainable.

4.3 Economic Concerns

The third panacea of sustainability, profit, found eminence in the supply chain, post 2010, by the increased attention of the management from competition between firms to competition between supply chains (Anderson and Skjoett-Larse 2009). Thus, the challenge is to '**effectively integrate the principles of sustainable supply chain practices into the existing supply chain programmes and operations**' of the focal company and also to '**ensure that sustainability considerations are embedded within all their sourcing processes**' (UNGC 2014).

The truth here is: value (and cost) is not just the culmination of the focal firm in a network, but also of all the different cross-connecting entities present in its eco-system; and this 'extended enterprise', as some have termed it, then, becomes a key factor of (gaining or losing) competitive advantage (Christopher 2012).

ITC Limited, an Indian multi-business conglomerate, recognizes the importance of extending its Triple Bottom Line (people, planet, profit) approach along its supply chain to ensure long-term competitiveness by adopting a balanced approach towards creation of social and environmental value and the economic viability of the businesses (ITC Limited 2012).

5 Managing the Foundation for Sustainability

Thus comes the concept of sustainable supply chain, as defined by Carter and Rogers (2008) as "*the strategic achievement and integration of an organisations' social, environmental and economic goals through the systematic coordination of key inter-organizational business processes to improve the long-term economic performance of the individual company and its value network.*" In this definition, one talks about two critical aspects: (a) strategic achievement; and (b) improvement of the long-term economic performance of the individual company and its value network. Or, in other words, to employ strategic thinking for long-term profit of the focal company and its value network, comprising of a larger number of firms with a more complex set of relationships between them, and agreements on a greater number of interface standards (Funk 2009).

Thus, ensuring sustainability in the supply chain, in this time of globalisation, where, it may be spread across multiple countries, different segments of buying agents, suppliers and subcontractors may be challenging. A company of **Wal Mart's** stature easily has 50,000–1,00,000 suppliers, spread all over the world, including India. According to the **Supply Chain Digest**, this retail giant stocks products made in more than 70 countries and at any given time, operates more than 11,000 stores in 27 countries around the world, and manages an average of 32 billion dollars in inventory (Lu 2014).

Moreover, as Vogel (2005) observed, the gap between the company and the industry codes and actual working conditions often remain large. This can be accounted to ineffective monitoring, independently unverifiable results and uneven compliance and violations (Vogel 2005); especially in a multi-tier supply chain. Therefore, monitoring and accountability needs to be done in an hierarchical process, one tier at a time till all the tiers are reached. Most often than not, majority of environmental, social and governance (ESG) risks lie deeper down the supply chain (Giblin 2013), resting with the grass root level suppliers, comprising mainly of the Micro, Small and Medium Enterprises (MSMEs).

In India, these MSMEs play a vital role for the growth of Indian economy. The annual report of MSME 2012–2013, has confirmed that the 44.7 million MSME enterprise with a total employment of over 100 million and more than 6000 quality products account for a large share of industrial units; as well as 43% of India's total exports in 2011–2012 (Ministry of Finance 2013). They act as the major link in the supply chain to corporate and the Public Sector Undertaking (PSUs) and significant contributors to (Ramesh 2013) large multinational enterprises (MNEs). By committing to engage in recognized traceability schemes, MSMEs can therefore demonstrate to customers their commitment to sustainability, and contribute to the achievement of MNEs, that increasingly demand more stringent sustainability requirements, and are looking for suppliers who can help them achieve these goals (UNGC 2014).

Therefore, sustainability is no longer the individual company's domain; and, as Anderson and Skjoett-Larse (2009) has noted, multinational companies are not only expected to behave socially responsible within their own juridical walls; they are also held responsible for environmental and labour practices of their global trading partners such as suppliers, third party logistics provider, and intermediaries over which they have no ownership, thereby impacting the world at large.

Thus, all companies, at different points, fall within the various tiers of the supply chain requirement. The large companies usually have the onus of managing large supply chains on one hand as well as comply as a supply chain to other companies; whereas Micro, Small and Medium Enterprises (MSMEs) also have the onus of complying as the supply chain of larger organisations (Indian and Multi National Enterprises), as well as act responsibly and manage its own suppliers (refer Exhibit 2). It is thereby, a multi-pronged/multi-tiered process.

To develop various efforts to make MSMEs more socially responsible by awareness raising and technological capability building, several attempts have been made by the government, the **Federation of Small and Medium Enterprises (FISME), World Assembly of Small and Medium Enterprises (WASME) and associations of importers in developed countries**, like the SUSBIZ India Project, initiated in 2007 by the Danish Federation of Small and Medium Sized Enterprises (DFSME) together with the Danish Commerce and Companies Agency (DCCA), that involved collaboration between eight Indian suppliers and their eight Danish

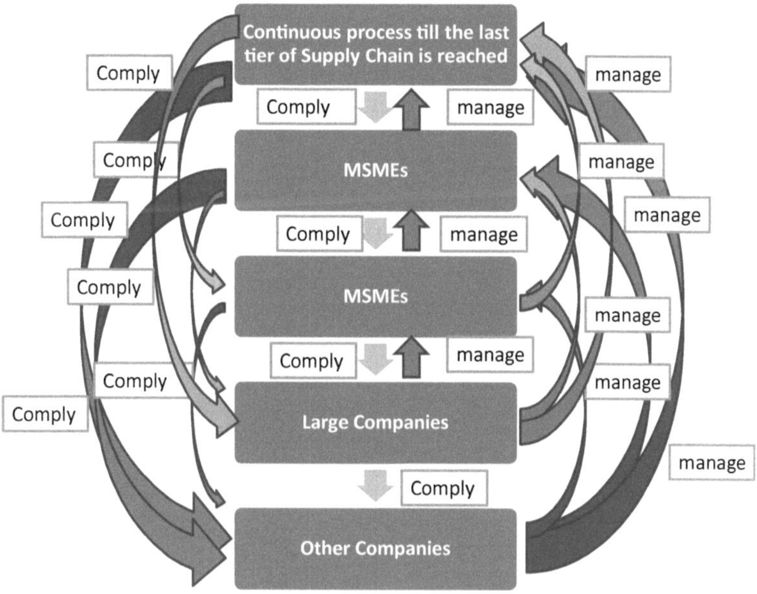

Exhibit 2 Dynamics of sustainable supply chain management

buyers (Sundar 2013). However, unless these compliances are percolated and adopted among the entire MSME sector, there will always be lacuna in the supply chain management and even the largest of the corporation will not be sustainable.

The **Indian government, in its National Voluntary Guidelines** on Social, Environmental and Economic Responsibilities of Business issued by the Ministry of Corporate Affairs, first in 2009 and updated in July 2011, realized that promotion of responsible business among the SMEs will need a multi-pronged approach, that includes "*facilitating SMEs to recognize the business case for adopting responsible business practices*" and "*preference by public agencies and large players in value chains to SME suppliers that follow CSR practices.*" (Sundar 2013).

In order to do this, the foremost requirement of the focal company is to be able to trace its supply chain till its last piece. It is extremely challenging, but not impossible. Let us take the example of Marico industries.

Saugata Gupta*, the Chief Executive Officer (CEO) of the fast moving consumer goods (FMCG) major, Marico Industries*, feels "*Supply chain is also a critical driver of efficiency in today's complex business environment, given the existing cost structures and challenges*" (Naren, 2013). Marico, the largest buyer of copra in India with a demand of about 100,000 tonnes a year, with the help of technology platforms could dis-intermediate its copra supply chain, in 2013, from the exploitative structures and agents, after a struggle for over two decades and trace it to the smallest possible vendor—a marginal farmer with a few coconut palms in his backyard from amongst 61 clusters involving 7982 farmers covering 1737 hectares for its popular range of Parachute coconut oils. This, indeed, is a very big successful transparency model (Naren 2013).

In view of the necessity of such deep rooted intervention in supply chain management and compliance, an entire cottage industry of auditors and consultants have emerged to advise manufacturers on the numerous regulatory efforts in place or on the horizon (Blanchard 2012).

The **United Nations Global Compact (UNGC)** also encourages participants to engage with suppliers around the ten principles (refer Table 2) and to advance

Table 2 The ten principles of the united nations global compact and supply chain sustainability

Human rights	
Principle 1: Businesses should support and respect the protection of internationally proclaimed human rights; and	Companies have a responsibility to respect human rights. The baseline responsibility is not to infringe on the rights of others. In addition, business can take steps to support and promote the realization of human rights, and there are good business reasons to do so
Principle 2: Make sure that they are not complicit in human rights abuses	
Labour	
Principle 3: Businesses should uphold the freedom of association and the effective recognition of the right to collective bargaining	Labour conditions in offices, in factories, on farms and at natural resource extraction sites such as mines, particularly in the developing world, often fall significantly below international standards and national regulatory requirements and can lead to serious human rights abuses. Businesses should strive to uphold international labour standards within their supply chains, including the right to freely chose employment, the freedom of children from labour, freedom from discrimination and the freedom of association and collective bargaining In addition, workers at times suffer from other labour rights abuses, including excessive work hours, degrading treatment by employers and inhibited movement. In order to avoid complicity in abuses, businesses should seek to ensure that they do not cause the rights of workers and others affected by their supply chain to be infringed upon, including the right to freedom of movement, freedom from inhumane treatment, the right to equal pay for equal work and the right to rest and leisure. The rights of all peoples to work in safe and healthy working conditions are critically important as well Companies can also begin to address human rights (including and beyond labour conditions) alone or by working with partners to promote a broad range of human rights such as gender equality and access to education and health
Principle 4: The elimination of all forms of forced and compulsory labour	
Principle 5: The effective abolition of child labour; and	
Principle 6: The elimination of discrimination in respect of employment and occupation	

(continued)

Table 2 (continued)

Environment	
Principle 7: Businesses should support a precautionary approach to environmental challenges Principle 8: Undertake initiatives to promote greater environmental responsibility; and Principle 9: Encourage the development and diffusion of environmentally friendly technologies	Environmental impacts from supply chains are often severe, particularly where environmental regulations are lax, price pressures are significant and natural resources are (or are perceived to be) abundant. These impacts can include toxic waste, water pollution, loss of biodiversity, deforestation, long term damage to ecosystems, hazardous air emissions as well as high greenhouse gas emissions and energy use. Companies should engage with suppliers to improve environmental impacts, by applying the precautionary approach, promoting greater environmental responsibility and the usage of clean technologies
Anti-Corruption	
Principle 10: Businesses should work against corruption in all its forms, including extortion and bribery	The significant corruption risks in the supply chain include procurement fraud and suppliers who engage in corrupt practices involving governments. The direct costs of this corruption are considerable, including product quality, but are often dwarfed by indirect costs related to management time and resources spent dealing with issues such as legal liability and damage to a company's reputation. Companies that engage with their supply chains through meaningful anti-corruption programmes can improve product quality, reduce fraud and related costs, enhance their reputations for honest business conduct, improve the environment for business and create a more sustainable platform for future growth

Source: http://www.bsr.org/reports/BSR_UNGC_SupplyChainReport.pdf

sustainable development objectives as part of their commitment to the Global Compact, and thereby to spread good corporate citizenship practices throughout the global business community (UNGC 2010).

6 Conclusion

Awareness on supply chain sustainability in the context of the developing/emerging countries like India, is still in the nascent state. It is a 'pull' and 'push' process, between the corporations and the Government, where each one thinks, it is the duty and responsibility of the other to make the rules and monitor the compliance. The debate is inconclusive. In reality, unless the sustainability norms are complied, in all its tiers of upward integration, that is the supply chain, the focal Company remains weak at its foundation and will have a tendency to collapse anytime like a 'house of cards.' This issue of supply chain sustainability needs to be looked into, as an investment rather than an expenditure by the corporation.

Kris Gopalakrishnan, *CEO and Co-Founder of Infosys,* an Indian multinational corporation, thus pointed out that "We live in an increasingly resource aware and resource constrained world. We need to live within our means and not borrow from the future. To build a sustainable tomorrow we need to make our supply chain sustainable today. In fact, I firmly believe that increased sustainability in the supply chain reduces risks and increases profits for all organizations and stakeholders" (UNGC 2010).

However, discussion on who makes the rules—the Government or the Corporations, or, maybe a multilateral agency like the United Nations to take care of the complex issues of supply chain sustainability, spread over different countries, cultures, local laws, could be the content of a separate study.

References

Anderson, M., & Skjoett-Larse, T. (2009). Corporate social responsibility in global supply chains. *Supply Chain Management: An International Journal, 14*(2), 75–86.

Blanchard, D. (2012, May 14). Corporate social responsibility in the supply chain: Manufacturers are expected to fully disclose where they source their products, or suffer the PR backlash. *Industry Week.* Accessed June 3, 2015, from http://www.industryweek.com/environment/corporate-social-responsibility-supply-chain

Carter, C. R., & Rogers, D. S. (2008). A framework of sustainable supply chain management: Moving toward new theory. *International Journal of Physical Distribution & Logistics Management, 38* (5), 360–387.

Christopher, M. (2012). *Logistics and supply chain management* (4th ed.). Harlow: Pearson.

Clough, J. (2015, January 9). Made in India: Five companies demonstrate global leadership for the climate. *Greenfutures Magazine.* Accessed June 3, 2015, from http://www.forumforthefuture.org/greenfutures/articles/made-india-five-companies-demonstrate-global-leadership-climate-0

Daily Mail Reporter. (2008, June 23). Exposed: Primark's fashion sweatshops that pay children just 60p a day. *Mail Online.* Accessed June 2, 2015, from http://www.dailymail.co.uk/news/article-1028449/Exposed-Primarks-sweatshops-pay-children-just-60p-day.html

Funk, J. L. (2009). The emerging value network in the mobile phone industry: The case of Japan and its implications for the rest of the world. *Telecommunications Policy, 33*(1–2), 4–18. Accessed June 3, 2015, from http://www.sciencedirect.com/science/article/pii/S0308596108000931

Giblin, C. (2013, November 7). Time for a multi-tiered approach to supply chain risk. *Sedex.* Accessed 5 June 2015, from http://www.sedexglobal.com/time-for-a-multi-tiered-approach-to-supply-chain-risk/

Gilmore, D. (2015, March 12). Any limits to large companies' supply chain control? *Supply Chain Digest.* Accessed June 2, 2015, from http://www.scdigest.com/assets/firstthoughts/15-03-12.php?cid=9092

Global Climate Change: Vital Signs of the Planet. (2015). *A blanket around the earth.* Accessed 2 June 2015, from http://climate.nasa.gov/causes/

ITC Limited. (2012). Sustainability report 2012: Key impacts, risks and opportunities: Sustainability challenges. Accessed June 1, 2015, from http://www.itcportal.com/sustainability/sustainability-report-2012/key-impacts-05.aspx

Lu, C. (2014, May 8). *Case studies, insights & analysis: Incredibly successful supply chain management: How does walmart do it?* Singapore: Tradegecko. Accessed June 5, 2015, from http://www.tradegecko.com/blog/incredibly-successful-supply-chain-management-walmart

McDougall, D. (2007, October 28). Indian 'slave' children found making low-cost clothes destined for gap. *The Observer, The Gurdian.* Accessed June 1, 2015, from http://www.theguardian.com/world/2007/oct/28/ethicalbusiness.retail

Ministry of Finance. (2013). Government of India. Report of the Inter-Ministerial committee for boosting exports from MSME sector. Accessed June 1, 2015, from http://dgft.gov.in/exim/2000/imc-EXPORT-sme.pdf

Naren, K. (2013, November 26). Sustainability & India Inc: Why marico's supply chain management is a case study at IIM-A. *The Economic Times.* Accessed June 3, 2015, from http://articles.economictimes.indiatimes.com/2013-11-26/news/44486941_1_saugata-gupta-fmcg-major-marico-supply-chain

Ramesh, M. (2013). Importance of worker engagement and empowerment for SME sustainability. In S. Mittal (Ed.), *CSR and competitiveness: Essential of the corporate India and its sustainability* (pp. 105–112). New Delhi: Allied Publishers Private Limited.

Sundar, P. (2013). *Business & community: The story of corporate social responsibility in India.* New Delhi: Sage.

Supply Chain Report 2014–15. (2015). Supply chain sustainability revealed: A country comparison: Report written for CDP by accenturestrategy. Accessed June 2, 2015, from http://www.accenture.com/SiteCollectionDocuments/PDF/Accenture-CDP-Supply-Chain-Report-2015.pdf

UNGC. (2010). Supply chain sustainability: A practical guide for continuous improvement. Accessed June 2, 2015, from http://www.bsr.org/reports/BSR_UNGC_SupplyChainReport.pdf

UNGC. (2014). A guide to traceability: A practical approach to advance sustainability in global supply chains. Accessed June 1, 2015, from https://www.unglobalcompact.org/docs/issues_doc/supply_chain/Traceability/Guide_to_Traceability.pdf

Vogel, D. (2005). *The market for virtue: The potential and limits of corporate social responsibility* (1st ed.). Washington, DC: The Brookings Institution.

Wisner, J. D., Tan, K. H., & Leong, G. K. (2011). *Principles of supply chain management: A balanced approach* (3rd ed.). Mason, OH: South Western.

Risky Business? On the Interplay Between Social, Actuarial and Political Risks and Licences

Martin Brueckner, Sara Bice, and Christof Pforr

1 Introduction

The last 15 years saw the rise to prominence of the concepts of 'corporate social responsibility' (CSR) and 'social licence to operate' (SLO).[1] The so-called business case for 'doing business the right way' has received considerable attention on the side of theory (Crane and Matten 2007). Arguably, it has also been instrumental in attracting business to the CSR and SLO realms where there has also been a growing realisation of the increasing cost to companies of paying insufficient attention to what were traditionally seen to be non-core areas of business; particularly among companies in the mining and extractives (M&E) sector with high social and environmental impact potential (Davis and Franks 2014; Willis 2013).

Today's M&E industry is highly, formally regulated and is governed by voluntary codes and corporate responsibility commitments (e.g. MMSD Project 2002). These include government regulations, such as environmental impact assessment (EIA) requirements, industry-based initiatives, especially the International Council

[1]For clarity, the acronym 'SLO' is used henceforth as a reference to SLO theory while the term 'social licence to operate' denotes the actual licence.

M. Brueckner (✉)
Centre for Responsible Citizenship and Sustainability, Murdoch University, Perth, WA, Australia
e-mail: m.brueckner@murdoch.edu.au

S. Bice
Melbourne School of Government, The University of Melbourne, Melbourne, VIC, Australia
e-mail: sara.bice@unimelb.edu.au

C. Pforr
School of Marketing, Curtin University, Perth, WA, Australia
e-mail: christof.pforr@cbs.com

H. Lu et al. (eds.), *Building New Bridges Between Business and Society*, CSR, Sustainability, Ethics & Governance, DOI 10.1007/978-3-319-63561-3_3

on Mining and Metals (ICMM) 'Principles for Sustainable Development' (ICMM 2003) and intergovernmental agency commitments, including the United Nations Global Compact (UNGC) (United Nations 2000). M&E firms are progressively expected to meet not only mandated regulatory requirements but also to demonstrate responsible behaviour towards the communities in which they operate (Dashwood 2012; Yakovleva 2005). Governments' expectations also stretch beyond regulatory compliance and include concern for companies' contributions to local and national economies and communities.

For many years, these expectations and related activities have been captured under the banner of 'corporate social responsibility' (CSR), 'sustainable development' or similar agendas. Lately, however, the 'social licence to operate' (SLO) concept has been gaining prominence (Thomson and Boutilier 2011) and taken a steadily important position in M&E companies' strategies and in communities' relationships with local operators (see for example Rio Tinto 2013). Communities are also using the term, largely as a means of resistance to unwanted M&E operations in their neighbourhoods (Bice 2014) or as a form of 'negative governance' (Owen and Kemp 2013). While both CSR and SLO are described as concepts with staying power and significant potential to influence corporate-community relations and activities (e.g. Bice and Moffat 2014; Holme 2010; Jackson 2014; Lacey and Lamont 2014; Moffat and Zhang 2014; Prno 2013; Prno and Slocombe 2012), they have also been subject to considerable criticism.

Critics suggest, for example, that far from encouraging public debate about impacts of community concern, the social licence rhetoric is used primarily as "an industry response to opposition and a mechanism to ensure the viability of the sector" (Owen and Kemp 2013: 29). Similarly, within the CSR space the dominant business case logic is seen to be limited to firms merely demonstrating their economic contribution (i.e. employment, taxes, royalties) to stakeholders, treating as implied their social licence to operate on the basis of the monetary benefits their operations promise to generate (Brueckner et al. 2014). The over-optimistic assumption of the frictionless convergence of commercial and social interests not only strikes as ideological (Toft 2015) and reminiscent of the Friedmanite position on CSR (Friedman 1970), it also risks reducing the SLO and CSR agendas to firms' economic legitimacy and fails to address the more deep-seated and potentially conflicted issues surrounding business-community relations (Brueckner et al. 2014; Brueckner and Mamun 2010).

In Australia—the focal point of this paper—cases of conflict between M&E companies and their host communities remain common (e.g. Brueckner and Ross 2010; Pini et al. 2010; Presten 2014; Wesley and MacCallum 2014). Despite improved corporate attention to CSR (Franks et al. 2009) and widespread espousal of the social licence (Bice 2014), companies' social legitimacy or that of their operations continues to be questioned by local communities even though projects and industries can garner political support and regulatory approval. Thus, far from offering a clear indication of good corporate behaviour, ongoing community opposition attests to a conflicted social licence terrain and raises new questions about the nature of the SLO and CSR concepts and their ability to capture, represent or explain the variety of risks and pressures exerted on M&E firms, affected communities and governments.

Reflecting on the current state of SLO and CSR in the Australian M&E industry, we argue in this paper that a broader conceptualisation of the risk and licensing terrain is needed to fully capture the nature and dynamics of corporate-community-government relations. To this end, the paper introduces what we call the SAP-troika of social, actuarial and political licences placed within a dynamic, comprehensive risk framework (adopted from Haines 2011) as a means of conceptualising a much wider SLO/CSR landscape perceived not just through the corporate lens but through that of all stakeholders. This conceptualisation contributes to scholarly research on SLO and CSR through its exploration of the concepts' 'politicality' and power dimensions and new contemporary theorisations of dynamic risks, especially social risk.

In what follows, we offer evidence of this situation and explore several central questions:

- What risks and licences constitute the SLO/CSR terrain?
- Of what nature is the relationship between the different risks and licences and how can this be captured in a single framework?
- How might such a framework help interrogate the meaning, intersections and probable implications of the various risks and licences theorised?
- What are the implications of such a framework for both SLO/CSR theory and praxis?

We commence with a review of recent developments in the theorisation of SLO and related discussions about CSR, especially with reference to the frequently downplayed issues of power and politicality. We then introduce and juxtapose the suite of social, actuarial and political licences and place them in Haines' (2011) dynamic risk framework with a view to explain the diverse and often competing concerns and dynamics among SLO/CSR stakeholder groups. The ensuing discussion will then focus on the consequences for this dynamic risk framework for understanding the set of licences required by today's M&E industry, and consideration is given in closing to opportunities for future research through employing the framework.

2 Conflicted SLO/CSR Theory and Praxis

The term 'social licence to operate' was coined in 1997 by ex-Placer Dome executive Jim Cooney, and the idea's wide acceptance in today's M&E industry and beyond (see Hall et al. 2013, 2015; Hall 2014; Lacey et al. 2014; Williams et al. 2007) led Cooney to quip recently that he wished he had trademarked the term (Leyne 2014). While social licences have remained a metaphorical device (Bice 2014), SLO expertise is increasingly being sought within industry to assist with the measurement and monitoring of firms' social licences (Prno and Slocombe 2012).

In recent years, growing and intensifying business-community conflicts over industrial development (see Bebbington et al. 2008; Klein 2007) have resulted in development proponents, communities and governments to place greater emphasis on the "power, role and expectations" of social licences and been a catalyst for research into their issuance and governance and the various procedural aspects of social licensing (Bice and Moffat 2014: 257). While still nascent a concept, as suggested by Bice and Moffat (2014), SLO has become ubiquitous and an extensively used means through which firms seek legitimacy by showing concern for social and environmental issues (Aguilera et al. 2007) yet without the pressures of binding regulation (Owen and Kemp 2013).

Whilst recognising the pitfalls of regulating social licences (Lacey 2013), the very notion of a social licence providing company or project legitimacy in the absence of compliance pressures (see Vogel 2008; Wilburn and Wilburn 2011) has also attracted critique (Owen and Kemp 2013). For example, the generation of the term from within the M&E industry, and proponents' tendencies to lay claims to a social licence without adequate stakeholder consultation and with limited reference to licensing criteria or brokerage (Bice 2014; Parsons and Moffat 2014) are found to be problematic. While the industry's acknowledgement of the importance of obtaining and retaining a social licence is welcomed, it is also criticised due to the largely naïve (or expedient) treatment of the inherent complexities associated with social licensing relating to the actual issuance of a social licence and attendant questions about informed consent, stakeholder inclusion and power relations (e.g. Solomon et al. 2008). As the SLO field remains theoretically and conceptually underdeveloped in this regard (Owen and Kemp 2013), industry practices often assume the form of a 'cost–benefit' or 'offset' approach to community relations, which rests on the assumption that adverse industry impacts can be adequately compensated for (Harvey and Bice 2014). Yet, especially for a sector that understands itself as a "development industry that creates new social possibilities" (Cutifani 2013) it is important for resource companies to engage with the 'local' and 'social' intricacies of their host communities (Owen and Kemp 2013) and establish relationships built on trust (Harvey and Bice 2014; Moffat and Zhang 2014). Instead of employing a risk and reputation driven approach that prioritises social issues primarily as a means of realising business benefits (e.g. minimal project delays) and merely responds to the business case of securing a social licence (Owen and Kemp 2013), firms need to develop an understanding of local contexts and the needs and aspirations of local people, demonstrating that 'they are listening' and 'keeping their promises' (Boutilier 2009). As suggested by Black (2013: 19) "benefits like bringing new jobs to a region [. . .] are often cited by companies as if this were enough to establish a social licence. It's not."

In contrast to SLO, early CSR discourse dates back to the early years of the twentieth century. Back then, it emerged as a counterweight to the corporate profit leitmotif and market libertarianism (Clark 1916), evolving into an altruistic and moral concept in the 1950s (Bowen 1953) while still remaining on the margins of academic debate and business practice. Only recent decades saw CSR shake its image of being a subversive doctrine (Lee 2008) and surface as a widely accepted basis for business success (Idemudia 2009). The now prominent 'business case for CSR' (Schreck 2011), which translates firms' attention to their social and environmental management obligations into business benefits (Branco and Rodrigues 2006; Hart 1997; Kotler and Lee 2005; Lantos 2001; McWilliams and Siegel 2001; O'Riordan and Fairbrass 2008; Porter and Kramer 2002, 2006, 2007; Schreck 2011) is underpinned by a 'win-win' ideology that is purported to overcome the profits-ethics dualism. Accordingly, companies can be profitable whilst contributing to the welfare of communities in which they operate (Banerjee 2007; Schreck 2011), a notion supported by research focused on the, albeit often tenuous, links between firms' positive financial and economic performance and their CSR strategies (Igalens and Gond 2005; Orltizky 2005; Schreck 2011; Vial 2007).

Not only does CSR provide corporate capitalism with a 'friendlier face' (Doane and Abasta-Vilaplana 2005: 23), it is also presented as a concept that is both 'rational' (Dean 2010: 18) and 'ideationally neutral' (Blowfield and Dolan 2008: 2). Yet, the very form and meaning of CSR remains highly contested, and CSR theory is far from being unified or coherent (Crane et al. 2008). Also, mainstream CSR scholarship does not acknowledge the complex, ideologically charged and politically contested nature of CSR (Blowfield and Frynas 2005; Idemudia 2010; Shamir 2010). The impact of contextual factors on CSR practice is a case in point (Argandoña and von Weltzien Hoivik 2009) with a growing body of literature pointing to marked differences in CSR governance and practice in different country and industry contexts (e.g. Geppert et al. 2006; Tempel and Walgenbach 2007; Turkina et al. 2015). In Australia's M&E context, for example, a number of industry community-conflicts surrounding resource development projects speak to the impact of the political environment on CSR practice, showing how a political pro-development climate can affect the nature and extent of resource companies' CSR strategies (e.g. Brueckner and Mamun 2010; Wesley 2014). Despite a growing awareness of how firms' social and political contexts affect corporate choices (Matten and Moon 2008) and shape their willingness to commit to, and operationalise CSR (Detomasi 2008), considerations such as these rarely feature in the mainstream CSR debate (Amaeshi and Amao 2009; Jackson and Apostolakou 2010).

The brief review above speaks to gaps within SLO/CSR theorising as well as a so-called 'decoupling' (after Meyer and Rowan 1977: 340) between SLO/CSR principles and practices (Bice 2015) with both fields at risk of being self-referential (Owen and Kemp 2013: 31) and self-declaratory (Morrison 2014: 141). While they highlight the importance of a genuine engagement with the social dimension of business and espousing a social licence, this engagement remains limited to the narrow logic of the business case of self-defined 'responsible practice', striking as an attempt at 'socialising' the economic mind-set of business (after Banerjee 2006) without challenging its fundamental values (Blowfield 2005). It is in this context that we argue for a shift away from the narrow focus in the SLO/CSR field and call (following Wesley 2014) for greater attention to be paid to the impact of power and politics as well as the structural and ideological influences that shape, define and limit the SLO/CSR agenda. Such a reorientation ought to entail a renewed focus on the 'social' within SLO and CSR theorising with greater emphasis on how, by whom and on what/whose terms various competing stakeholder interests are balanced and addressed. The SAP framework introduced below is intended as a vehicle for the explication of these conflicted aspects of SLO and CSR.

3 The SAP-Troika

In this section we introduce and juxtapose social, actuarial and political licences, following a similar approach taken by Morrison (2014), and then embed this set of licences in the dynamic risk framework developed by Haines (2009, 2011). In doing so, we combine key risk and licence requirements to explore of how these are addressed and weighted by all SLO/CSR stakeholders. Specifically, the model helps illuminate the tensions inherent in the SLO/CSR space and capture the meaning, intention and probable implications of the various risks and licences at play.

3.1 Social Licence to Operate

While definitions vary, a social licence to operate is generally understood to be: "the ongoing acceptance and approval of a [project] by local community members and other stakeholders that can affect its profitability" (Moffat and Zhang 2014: 61). Requirements for such a licence run the gamut from worker safety to cultural sensitivity, and the degree of social licence granted by a community may range from withheld/withdrawn through to assimilation of a firm within the community fabric (Thomson and Boutilier 2011).

Social licences are not formally issued and thus cannot be seen as tangible contracts or documents (Franks and Cohen 2012). Instead, they are better understood as a form of social acceptance approval by companies through trust building engagements with stakeholders (Moffat and Zhang 2014) and meeting local community expectations as well as those of the wider society and other constituents (Gunningham et al. 2004). This forms the basis for the legitimacy of a company or a project (Black 2013: 15). According to Morrison (2014), legitimacy, trust and consent are preconditions for the existence of social licences, which are also affected by contextual factors such as the creation of wider societal benefits, the sharing of knowledge and relinquishing of power and providing transparency and accountability as well as minimising and adequately compensating for adverse impacts (see Fig. 1). The corporate quest for a social licence centres on relationship building with stakeholders and thus differs from philanthropic spending and investments in communities (Morrison 2014).

The issuance of a social license often necessitates that companies go beyond mandatory, regulatory compliance (Gunningham et al. 2004) and address growing societal expectations that stretch to the political and social realms (Bice 2014;

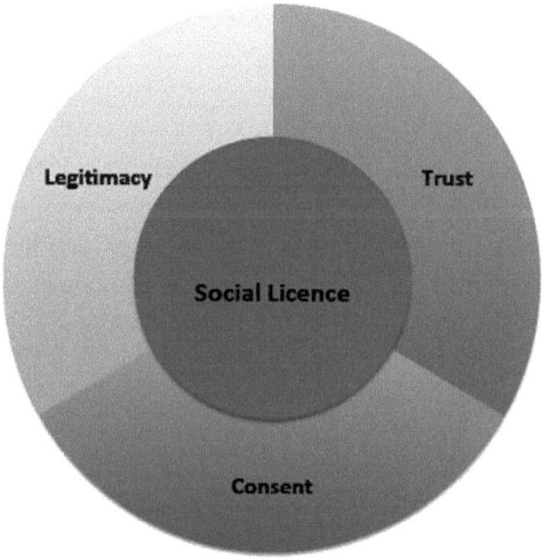

Fig. 1 The social licence and its building blocks

Minerals Council of Australia (MCA) 2013; Schmidpeter 2013). The wide uptake of this broader CSR agenda is said to be evident in the M&E sector despite the constraining character of CSR on company activities (Dashwood 2012). Far from constraining, however, certain CSR and social licence related activities may be seen as stretching M&E companies beyond appropriate roles in communities and even exonerating governments of traditional responsibilities (Harvey and Bice 2014). Then again, the adoption of an enlightened SLO and CSR rhetoric can also be seen as a form of 'discursive regulation' (Bridge and McManus 2000: 22) and as a way of minimising regulatory impositions by government (Hamann and Acutt 2003; Parsons et al. 2014).

Recently, Jijelava and Vanclay (2014) suggested that multiple, competing social licences may exist within one community or relative to one project. Here, the authors see various social licences as tied to particular stakeholder groups—in this instance, a gendered social licence—which requires project proponents to "obtain the approval and active engagement of all their stakeholders", each of whom may have different licensing requirements and, therefore, different social licences (Jijelava and Vanclay 2014: 288). While intriguing, this early conceptualisation perhaps raises more questions than it answers, and the proposition seems problematical. For example, who grants or withdraws the various licences? How is one social licence more or less valuable than another? In the end, which social licence is accepted, which rejected? Or are multiple social licences possible or required? While flawed, this notion raises ideas very helpful for reconceptualising the requirements for successful, contemporary M&E operations.

3.2 Actuarial (Legal) Licence to Operate

Most organisation types require legal or actuarial licences granted by a government authority so as to be able to operate lawfully; this includes for-profit and not-for-profit entities (Morrison 2014). These licences refer to permits and approvals enshrined in regulation, stipulating minimum standards and requirements to be met by organisations and rules of conduct to be followed. Examples of such standards include environmental licences, emission permits, project approvals or occupational health and safety standards.

In the 1960s, the attainment of high levels of economic prosperity in developed countries with unrestricted access to resources was largely unhindered by regulation. However, the side effects of economic activity triggered concern in light of increasingly visible impacts on social and environmental well-being (e.g. air and water pollution, hazardous waste) (Eckersley 1998; Jacobs 1991). Government regulation was considered necessary to restrict economic activity to contain the fallout, giving rise to policies that sought to impose constraints on industry (Panayotou 1994). Since the advent of more exacting government regulation, legal compliance pressure on business has increased especially in high impact industries such as the M&E sector where governments in numerous countries have started to regulate more closely the planning, operation and closure of mines (Bereton 2002).

In this sense, actuarial licences can be reflective (at least of aspects) of social licences for they give legal standing to what are common societal concerns such as process or product safety, emission controls or working conditions. Regulation in this sense can be seen as means of reducing friction between economic and social interests (Bridge and McManus 2000) and correcting markets failing to achieve public interest (Bomsel et al. 1996). The regulatory space, however, is highly contested. Not only, as will be explored further below, are the forces of globalisation and the rise of corporate power as well as the ever-present fear of government failure and shrinking role of the state challenges to regulation and its effectiveness (Matten and Crane 2005; Scherer and Palazzo 2011), regulators also need to find a balance between social and economic demands and are answerable to the public, media, and non-government organisations (NGOs) as well as the business community and special interest groups (Morrison 2014); thus, there are considerable political risks and trade-offs involved in policy formulation. In the M&E sector in particular, there is considerable tension between demands on the one hand to protect social and cultural values as well as community health and well-being whilst on the other hand there is pressure to cut 'red and green tape' as a means of fast-tracking resource development approvals and maintaining industry competitiveness.

Such tensions are exemplified by ongoing conflicts in Australia, for example, where communities are withholding a social licence from the country's nascent 'unconventional' gas industry. In Western Australia (WA) alone over 60 local councils have in recent years proactively withdrawn a social licence from the industry prior to the development of local gas resources while in other parts of the country communities have declared themselves 'gas-field free' in an attempt to prevent the establishment and expansion of gas wells and hydraulic fracturing ('fracking') processes (see Conservation Council of Western Australia 2014a, b; Gasfield Free Northern Rivers 2015; Hadji and Sweeney 2014). Such community opposition is occurring despite these projects and industries garnering political support and even receiving regulatory approval, affirming Bruce Harvey's view (cited in Morrison 2014: 15), that "a piece of paper from a government authority is often not enough for an activity to proceed". Situations such as these highlight the significance of a social licence and help explain industry investments in beyond-compliance measures (Gunningham et al. 2004).

3.3 Political Licence to Operate

Recent research asserts that the SLO concept is no longer adequate to explain the competing interests, values and agendas influencing M&E firms' behaviours in relation to their social and environmental impacts and choices (Brueckner et al. 2014). Such concern has led to theorising about the real-existence of a 'political licence to operate'. Based on an in-depth study on the sustainability of mining in WA (Brueckner et al. 2014), Brueckner et al. (2014) found that the state's 'developmentalist' policy agenda (after Kellow and Niemeyer 1999) overran social licence concerns related with the WA government prizing and sanctioning 'economic legitimacy' (after Boutilier and Thomson 2011). In the face of localised community agitation against mining interests both the mining industry and the government were seen to place greater emphasis on the resource sector's contribution to the state's economic agenda. Local concerns were traded off against the industry's contribution to investment, employment creation and royalty payments. As a result, the political licences issued by the WA state government to mining interests were found to have subordinated and masked the very social and environmental impacts that would normally affect the issuance of an earned social licence. Perceived weaknesses in state regulation, monitoring and enforcement (see Chandler 2014; Roche and Mudd 2014) served to compound what Bice (2013) calls 'crises of identity and sustainability', referring to the plight of communities most affected by resource development and thus most likely to benefit from a social licence but "often forgotten, frequently misunderstood, and [...] comparatively less researched" (Bice 2013: 138).

Based on the above findings the authors defined the political licence as "a politically derived licence representing government approval of, and support for, an industry based on its contribution to a state's economic development agenda" (Brueckner et al. 2014: 315). For the purposes of the model introduced in this paper, however, a broadening of this perspective is needed and to be understood more widely as a licence applicable to the legitimacy of government as well. More generally, a political licence might be thought of as both a licence to govern and an authority given by government to an organisation to undertake a particular activity (Morrison 2014: 21). Firms economic contributions may certainly be a factor in the issuance of a political licence, yet their contribution to the public interest also matters for political licences also include social licence aspects. While in authoritarian regimes a political licence may be all that matters (Morrison 2014: 21), in democratic settings political decision making ought to be aligned with broader societal interests as misalignment can translate into political risks and voter backlash. To illustrate, in the early 2000s there was considerable public disquiet in WA about what was known as the Coral Coast Marina Development, which envisaged developing a 2500-bed resort-style marina at Mauds Landing situated near the World Heritage listed Ningaloo Marine Park. Even though the project had obtained two approvals by the state's Environmental Protection Authority (EPA), in the face of considerable local, national and international protest as part of the 'Save Ningaloo' campaign (Gilles et al. 2004) the WA state government finally rejected the proposal (Gallop 2003). Whilst citing unacceptable environmental risks in its decision despite existing EPA approvals, the government is likely to have acted on the basis of perceived political risks (Brueckner and Pforr 2011).

3.4 Risk and Regulation

The widespread acceptance of the social licence and the increasing influence of the political licence point to the tight interlacing of voluntary and mandatory regulation for impactful industries. While our aim is not to engage in debates about the most appropriate voluntary/mandatory regulatory mix or to suggest improvements to regulatory practice, the connection between regulation and our focus on risk and licences cannot be ignored. At least for the M&E industry, this discussion is inseparable from current attitudes and approaches to regulation, and it is worthwhile setting out this context before detailing our 'social, actuarial, political (SAP) licence and risk model'.

There is currently a global trend for governments to pare back regulation in the quest not to *de*regulate, but to create 'good regulation' (Aguilera et al. 2007; Matten and Crane 2005). This movement can be seen in the development of new regulatory tools and improved regulatory efficiency (Haines 2011), and in widespread, post-GFC efforts to avoid 'regulatory capture'—wherein regulators become too close to regulated entities, compromising enforcement (OECD 2013). It is also reflective of shifts in attitudes about what regulation can and cannot achieve for businesses and communities. It represents decreasing government control over multinational corporations that themselves have become more active political actors (Scherer and Palazzo 2011), a circumstance that has partly influenced the proliferation of transnational governance initiatives, especially in the M&E industry. From a business perspective, previously accepted arguments that business gains from regulation (Stigler 1971) have largely been replaced by a neoliberal deferral to the market (Grabosky 2013) and a sense that regulation can also be counterproductive

(Grabosky 1995). Alternatively, scholars like Christine Parker (2002) argue that it is the changing nature of the corporation and the institutionalisation of responsibility which drives increased corporate 'self-regulation'. Still others assert that communities empowered with new media skills and the ability to hold corporations to account result in progressively stringent expectations and standards, especially in hazardous or highly impactful industries (Braithwaite 2008).

In Australia, for example, the acceptance of the social licence in the M&E industry is strongly linked with desires to reduce 'red' and 'green' tape while simultaneously promoting best practice. The tape cutting agenda is one fully supported by peak industry bodies, including the Minerals Council of Australia (MCA 2014) and the Australian Petroleum Production and Exploration Association (APPEA 2014). The Commonwealth Government's Productivity Commission recently launched a 'cutting red tape' website and is pushing a 'one-stop-shop' for environmental licenses and a broad red tape-cutting agenda (Australian Government 2015). Meanwhile, the Chief's Scientist's Office explicitly links M&E industry activities with the national economy, environment and community benefit in its 'Science and Research Priorities for the Nation' (Chubb 2015), underlining the importance of achieving an effective and efficient regulatory mix.

The push for tape-cutting, combined with the other drivers highlighted above, has contributed to an acceptance of governance beyond government, a situation where good regulation is not necessarily government-derived or mandatory and where voluntary or self-imposed regulations proliferate (Vogel 2008) and firms seek to perform 'beyond compliance' (Gunningham et al. 2004). This environment supports acceptance of social and political licences as legitimate regulatory mechanisms. The criteria for good regulation, then, are not enshrinement in legislation, but regimes marked by the qualities of: accessibility, accountability, consistency, transparency, being well-targeted at the concern in question, effectiveness in reducing risk and enforceability (Haines 2011). We propose that a dynamic risk perspective offers a helpful and pertinent means of understanding the increasing acceptance of governance mechanisms, especially the social and political licences to operate.

Growing acceptance of voluntary governance arrangements also demonstrates the progressively critical role that stakeholders beyond corporations and government play in how we conceptualise social and political licences and related risks. Indeed, community support for or objection to industries or projects is heavily informed by particular stakeholders' perceptions of risk (Haines 2009, 2011). The ways in which individuals *perceive* risk and harbour desires for regulation relate directly to their judgments about the types of risk and their potential impacts and benefits especially those which are most hazardous or even fatal (Slovic et al. 1984). Thus, particular stakeholders' values, priority concerns, risk perceptions and tolerance, and regulation (here via licensing) intersect. Attention to perceptions of risk allows us to conceptualise the full spectrum of risk types to avoid being reductionist and to prevent the weighting of one group of stakeholders' risks over another (e.g. actuarial vs. socio-cultural).

4 The Social, Actuarial, Political (SAP) Licence and Risk Model

Risk is, itself, a contested term, stretched in disparate directions by competing ontologies (Kaplan and Garrick 1981; Rosa 1998). Social scientists generally agree that risks are "determinable, calculable uncertainties" (Beck 1995: 77), but which cannot necessarily be "reduced to the following of scientific rules and procedures" (Rosa 1998: 20). As such, risk is political in nature (Sapolsky 1990). It is also dynamic and not irreducible to any one type. This does not mean, however, that risk is wholly subjective. Instead, as Haines suggests, "it is not only possible but necessary to measure and compare across the spectrum of risks facing society at any one time in order to prioritise resources" (Haines 2011: 32). Identifying the various types of risk through the dynamic risk framework detailed below helps to pinpoint and tease out competing interests or purposes (Dunn 2000) which inform conflict or cooperation around select issues or industries; here, the M&E industry. It allows for a weighing up of impacts and benefits (Guillaume and Charron 1999; Haines 2011; Yim and Vaganov 2003).

In the sections that follow, we present a holistic, social, actuarial, and political (SAP) licence and risk model that underpins the social, political and actuarial licences we argue are necessary for the successful operations of contemporary M&E firms. Thus, the framework presents 'three independent yet intersecting ideal types of risk' (Haines 2011: 34). Throughout the discussion, we explore how these diverse risk types correspond with actuarial, social and political licences.

4.1 Actuarial Risk

Actuarial risk is perhaps the most widely applied conceptualisation of risk, particularly among corporations in relation to auditing and corporate governance. It is traditionally associated with 'physical or financial threat' (Haines 2011: 34) where there is a 'reality of harm' (Haines 2011: 36). It is also commonly associated with 'business risk' (Graetz and Franks 2015). Here, risk most often equates to an expected loss, with measurable severity and predictability (Aven and Renn 2009). The International Standards Organisation (ISO) captures this perspective succinctly when it defines risk as the "effect of uncertainty on objectives" (International Standards Organisation 2009). It is this approach to actuarial risk as that which can be anticipated, built into scenarios, tested and (theoretically) prevented or mitigated (Kaplan and Garrick 1981), which results in it being the type of risk most commonly regulated. In the M&E industry, actuarial risk is thereby closely associated with exploration or project licences awarded by government.

For example, experience, data and comparability between operation sites facilitates creation of reliable risk frequency and probability scenarios concerning a proposed mine's environmental impact. This allows companies to anticipate and analyse 'acceptable risks' (Kaplan and Garrick 1981) and to make decisions about the environmental risks they are willing to bear, the risks which are acceptable under relevant regulation, and the resources and procedures which must be in place to prevent or mitigate particular risks to meet requirements and achieve project licensing. This is reflected in the close relationship between environmental impact assessment (EIA)—a risk and benefit assessment of operations—and the regulation of EIA in disparate locations globally (Morgan 2012). In line with actuarial risk's focus on probability and frequency (Kaplan and Garrick 1981), attention to actuarial risks through regulated EIA and incorporation into international law and global lending standards (e.g. Equator Principles) (Morgan 2012), reflects the perception of environmental risks as posing real but identifiable harm which can be predicted, measured and mitigated.

Through its close relationship to regulation, actuarial risk is positioned as a kind of threshold risk. It captures those issues—most commonly environmental and financial—that must be addressed in order to achieve minimum, government-based approvals (e.g. regulatory licences) for project operation. But as the contemporary governance environment described above suggests, attention to actuarial risk and related actuarial licensing are no longer sufficient to support successful M&E operations. Nor is it suggested that attention to actuarial risks, as they are defined here, is sufficient to cover off all risks associated with a project; hence, our consideration of social and political risks.

4.2 Social Risk

Socio-cultural risk (here, social risk) is most often presented in contrast, but also as a necessary complement, to actuarial risk. It is variously described as "a situation or event where something of human value (including humans themselves) is at stake and where the outcome is uncertain" (Rosa cited in Aven and Renn 2009: 1) or as "events or behaviours identified as a threat to social order by a particular community at a particular time" (Haines 2011: 43). The key to social risk is that it values *perceptions* about potential risks or hazards, thereby validating community concerns in a way actuarial risk cannot. While usually more value-laden, moral or emotional than those risks classed as actuarial, social risks are not inherently irrational. Social risks are sincerely experienced by individuals and communities as signs of danger and allow community members to signal their concerns for the broader society in which they live (Douglas 1992). Thus, attention to social risks may become an important means of performing citizenship and belonging in society, revealing "an emotional as much as an intellectual logic" (Haines 2011: 44).

For M&E operations, social risks demonstrate the complex interweaving of those issues (e.g. environmental impacts) which can be objectively measured and managed via actuarial risk assessment and the—oftentimes varying (Black 2013)—community perceptions and concerns which may be more difficult to pin down but are felt just as strongly. As the Integrated Risk Governance Council explains, "both the 'factual' (i.e. actuarial) and the 'socio-cultural' dimension of risk need to be considered if risk governance is to produce adequate decisions and results" (Integrated Risk Governance Council 2005: 12). Or, as Haines (2011: 45) describes it, "[Communities'] interdependence is centre stage and with that [their] interaction with government and other sources of authority are critical." This positioning speaks directly to the role of governance and the relevance of the social licence. When linked to social risk and when understood as being in dynamic play with actuarial and political risks (and their related licences), the granting of a social licence signifies an appropriate level of attention to and action against social risks.

It is important to note that the conceptualization of social risk embraced here is in contrast to traditional understandings of 'social risk' in the M&E industry, wherein social risk has been commonly defined as "the risk(s) to businesses/ operations arising from interactions with, and the actions of, host communities" (Graetz and Franks 2015: 2). Graetz and Franks (2015), for instance, have demonstrated the importance of teasing out social from business (i.e. actuarial) risk to avoid negative impacts for business and communities, especially in the M&E industry. It is equally important, therefore, that social risks are attended to through regulatory or governance mechanisms that acknowledge their uniqueness. In the absence of regulation, the social licence has an important role to play here. Yet even where attention is paid to social and actuarial licensing requirements, political pressures introduce a third type of risk, one which may play a particularly decisive role where actuarial risks can be addressed but social risks remain concerning.

4.3 Political Risk

Political risk comprises two major, sometimes competing concerns. First, it relates to the risk faced by an elected government of losing legitimacy and the commensurate political capital and authority which such legitimacy delivers (Haines 2011). Legitimacy-related political risks extend far deeper than the 'legitimation crises' (Habermas 1979) linked to re-election or concern with the 24-h news cycle, although these may be of immediate concern to governments at high-risk. Political risks hold the potential to undermine Government's legitimacy, with the worst-case situations resulting in conflict and even violence. Secondly, political risk relates to the economic responsibilities and accordant risks governments face via their responsibility to ensure appropriate resources to support the state.

In political risk, we see a tight interweaving of economic drivers and requirements—both local and global—and the ways in which economic performance may influence a government's legitimacy in the eyes of its constituents. Political risk is particularly knotty where economic pressures may encourage policy decisions, which—while promoting economic prosperity—fail to acknowledge or trade-off identified social or actuarial risks. The M&E industry in Australia again provides a salient example and demonstrates the ways in which political risks affect a government's political licence. In June 2010, following a heated public debate with the M&E industry over the Labor Commonwealth Government's proposed 'Resources Superprofits Tax', Prime Minister Kevin Rudd was ousted by his deputy, Julia Gillard. While more complex than a stoush over taxes, the influence of the M&E industry in Australia—the sector is regularly credited as the reason Australia 'survived' the GFC—played a major role in undermining Rudd's political legitimacy (Chubb 2014). Today, following another change in leadership and election of a Liberal-National Coalition Government, a leading Australian economist warns that the Commonwealth Government's short-term focus on the economic component of political risk brings their legitimacy into question, as difficult, long-term policy decisions are being avoided (Garnaut 2014). Thus, the competing pressures of political risk relate directly to a government's political licence. Or, as Douglas (1992) explains, the public holds a desire to account for political risk, which is reflected in a more basic desire for order and the normative constraints commensurate to that desire.

4.4 *The SAP Framework*

The above exploration of social, actuarial and political risks and licences spoke to the connections that exist between them. These relationships are captured in the SAP framework shown in Fig. 2 below. In this section, we seek to query these connections further and show how the various risks and licences are not only related but also in tension with one-another and may indeed be competing, depending on perspectives, agendas and the power of various stakeholders related to different types of risk.

Central to all three licences is a public interest component. While the social licence can be seen as a direct assertion, and expression of public interest, both actuarial and political licences are arguably intended to be reflective of it. While this may be seen as a theoretical ideal applicable only to representative democracies, it bears note that conflict arising over any of the three licences—as illustrated by examples further below—generally entails perceptions of compromised public interest and thus also has direct bearing on all three risk domains.

Social, actuarial and political risks are born out of the tensions between the social, actuarial and political licences displayed in the respective corners of the diagram and the risks the licences themselves entail. The three licences are connected and interdependent (Morrison 2014) but also in friction with one another. As suggested by Brueckner et al. (2014), for example, the Western Australian state government's strong developmentalist agenda effectively outweighed communities' withholding of a social licence from various M&E industry projects, privileging the government's political licence. However, while in situations where political licences are dominant, political or commercial interests may lay claim to a social licence in relation to a particular development project, the mere assertion of a social licence is

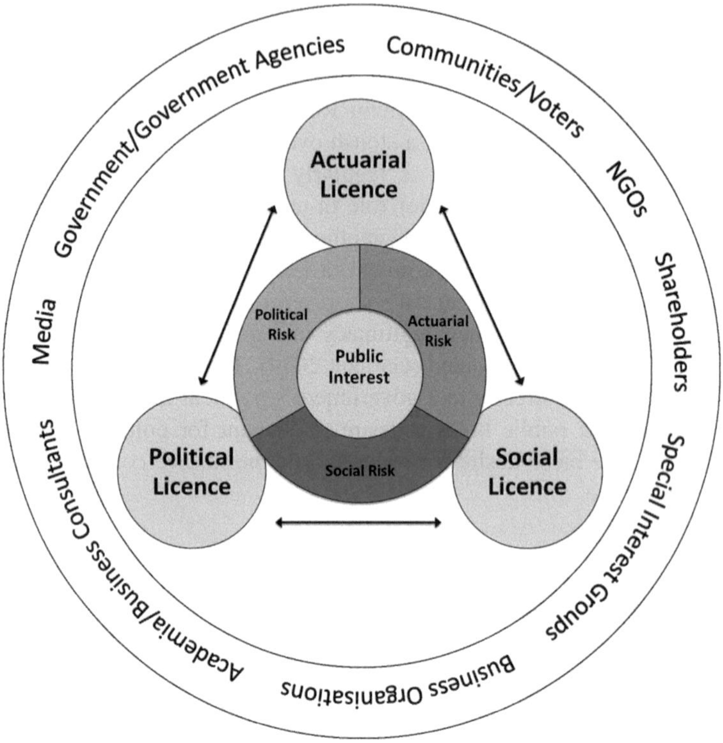

Fig. 2 The SAP framework. Sources: Based on, and adapted from, Haines (2009, 2011) and Morrison (2014)

open to challenges and can thus translate into considerable risks for those claiming to have it. For example, conflict erupting over coal seam gas (CSG) developments in Victoria and New South Wales in recent years saw public opposition to the 'supposed' issuance of both social and political licences, prompting the two states to impose temporary moratoria on 'fracking' operations. These suspensions were placed by both state governments even though in many of the conflict areas actuarial licences were obtained by the respective industry proponents (Grudnoff 2014; Organ 2014). In other words, challenges to social and political licences can override actuarial licences. In New South Wales and Victoria growing social and political risks arguably triggered this. Social risk on the one hand centered on public perceptions of unacceptable harm to human and environmental health and concerns about land rights and/or impacts on other industries (Hoare and Rose 2011). On the other hand, political risk grew due to increasingly widespread public opposition to the CSG industry, comprising of unlikely—albeit powerful—allies such as farmers, conservationists and traditional land owners (Kuch et al. 2013; Walker 2012). Also, actuarial licences can at times be seen to trump both political and social licences. Independent assessments carried out by statutory bodies, for example, under Environmental Protection Legislation can run counter to both political and/or social interests (Hasham 2014; Towie 2013). Yet, not only do ministerial powers generally allow for the potential side-lining of 'adverse' findings and determinations by an Environmental Protection Authority, rules and regulations—as shown in the ongoing conflict over coal mining in the Hunter Valley in New South Wales, which prompted a series of changes to rules governing project approvals (Hannam and Nicholls 2015)—are subject to change and themselves a source of conflict between interested stakeholder groups.

The cast of stakeholders weighing into debates about social, actuarial and political licences shown in the outer circle of the SAP model can be vast and varied depending on the context and issue domain. While governments, communities and private sector actors are generally prominent in such debates, licence negotiations or contestations frequently also involve the media, research organisations, lobby groups and NGOs. In certain contexts, the stakeholder network could even be widened to include actors at an international level, supranational law-makers and standard setting organisations as well as the investment community and social and environmental movements. Such a broad array of stakeholders, for example, can be found in the intensifying conflict over proposed coal mining operations in Queensland's Galileo Basin and associated infrastructure developments. In this context, the size of the stakeholder network is determined by the local, regional and global ramifications of the project related to its climate change impacts, associated risks to the World Heritage listed Great Barrier Reef, traditional land rights and local community concerns as well as international finance and foreign corporate interests (Robertson 2015; Taylor 2015).

With different stakeholder groups representing segmental interests and operating from different power bases, as shown previously, licence negotiations are prone to be highly complex and political in nature. Government and industry interests may pursue development goals and price economic values (e.g. CSG or coal in Australia), while community stakeholders may hold different value sets arguing for social justice, environmental sustainability or Indigenous rights (Solomon et al. 2008: 144). Independent expertise is often called upon to either diffuse tensions or to bolster positions held within licensing debates, frequently leading to the politicisation of science and expert knowledge in the competing claims arena (Beck 1986; Giddens 1990; Jasanoff 1986). The media in this regard can often, even unwittingly, act as an echo chamber for conflicts and inform as well as frame public perceptions and, in doing so, act politically but also become politicised (Gilding et al. 2012; Herman and Chomsky 2002).

Overall, it is this power-laden interplay of competing demands and interests that determines the level, and kinds of risk licence stakeholders face compounded by the competing risks the licences themselves pose. This was shown earlier in relation to political risks and the need for the careful balancing of a government's political legitimacy and its economic responsibilities. The SAP troika thus sheds more light on a highly dynamic licensing environment in which complex stakeholder arrangements and often colliding stakeholder interests necessitate the negotiation of licences to achieve an outcome that is in the public interest.

5 Conclusions

In this paper we hope to have highlighted what we consider conceptual shortcomings in SLO and CSR theorising concerning the social licence to operate concept but also to have offered a platform for bringing together these related yet disparate strands of theory. We have broadened the social licence spectrum by way of including political and actuarial licences to provide an overview of the wider conflicted licensing terrain and then placed the three licences into a dynamic risk framework to show each licence in context with its corresponding risks for different licence stakeholders. We have further illustrated how this multitude of at times aligned and other times competing interests shapes the risk profiles of actors within the social, actuarial and political licence space, in turn affecting stakeholder choices and behaviours and thus potentially licensing outcomes.

The SAP troika, to our reading, is a useful device for giving expression to the different motivations of licence stakeholders and explicating the inherent tensions between them. By making visible the generally complex stakeholder arrangements, underlying drivers and associated power dynamics, the model has a descriptive as well as predictive character in that an understanding of actors' motivations, their risk profiles and relative power can assist in projecting outcomes in licensing negotiations. The model also helps highlight the double-edged nature of licences, showing how licences subject to stakeholder constellations and power dynamics, can translate into risks which themselves are varied. The model's central focus on the public interest has also enabled the recalibration of what is largely a corporate risk perspective within mainstream SLO and CSR theory, emphasising the centrality of the 'social' in the licensing space.

The SAP troika makes plain that social licences are layered with risks and interwoven with other licences and therefore far more complex than the dominant social licence discourse might suggest. In this sense, we believe the model has application to SLO/CSR theory and praxis for making explicit these aspects and dimensions. Yet, moving forward, and understanding this model to be still at its early development stage, empirical testing is needed of the risk and licence relationships we have identified. Research of this nature would assist also in determining whether the suite of risks and licences included here are sufficient and able account, for example, for firms' financial risks, which actuarial risks may only partially cover. A similar broadening of the field could also be envisaged in relation to stakeholders to arrive at a more finely grained understanding of the various actors in the licensing space and their respective motivations and risk profiles. Importantly, such as research agenda would help develop further insights into the nature and workings of the social licence to operate which is central to the SLO/CSR enterprise, this paper's starting point. In this regard, through its conceptualisation of a much broader social licence context we believe the SAP troika provides a solid platform for further research.

Acknowledgements Parts of this chapter have been published in Bice, S., Brueckner, M. & Pforr, C. (2017). Putting social license to operate on the map: A social, actuarial and political risk and licensing model (SAP model). Resources Policy 53: 46–55 and are reproduced here with permission.

References

Aguilera, R. V., Rupp, D. E., Williams, C. A., & Ganapathi, J. (2007). Putting the S back in corporate social responsibility: A multi-level theory of social change in organizations. *Academy of Management Review, 32*(3), 836–863.

Amaeshi, K., & Amao, O. (2009). Corporate social responsibility in transnational spaces: Exploring influences of varieties of capitalism on expressions of corporate codes of conduct in Nigeria. *Journal of Business Ethics, 86*(2), 225–239.

Argandoña, A., & von Weltzien Hoivik, H. (2009). Corporate social responsibility: One size does not fit all. Collecting evidence from Europe. *Journal of Business Ethics, 89*(3), 221–234.

Australian Government. (2015). *Cutting red tape*. Accessed May 21, from https://cuttingredtape.gov.au

Australian Petroleum Production and Exploration Association. (2014). *Red tape repeal day*. Accessed 21 May, from http://www.appea.com.au/media_release/red-tape-repeal-day/

Aven, T., & Renn, O. (2009). On risk defined as an event where the outcome is uncertain. *Journal of Risk Research, 12*(1), 1–11. doi:10.1080/13669870802488883.

Banerjee, S. B. (2006). The problem with corporate social responsibility. In S. Clegg & C. Rhodes (Eds.), *Management ethics: What is to be done?* (pp. 55–76). New York: Palgrave.

Banerjee, S. B. (2007). *Corporate social responsibility: The good, the bad and the ugly*. Cheltenham: Edward Elgar.

Bebbington, A., Hinojosa, L., Bebbington, D. H., Burneo, M. L., & Warnaars, X. (2008). Contention and ambiguity: Mining and the possibilities of development. *Development and Change, 39*(6), 887–914. doi:10.1111/j.1467-7660.2008.00517.x.

Beck, U. (1986). *Risikogesellschaft auf dem Weg in eine andere Moderne*. Wiesbaden: Suhrkamp.

Beck, U. (1995). *Ecological politics in an age of risk*. Cambridge: Polity Press.

Bereton, D. (2002, September 3). *The role of self-regulation in improving corporate social performance: The case of the mining industry*. Australian Institute of Criminology Conference on Current Issues in Regulation: Enforcement and Compliance, Melbourne.

Bice, S. (2013). No more sun shades, please: Experiences of corporate social responsibility in remote Australian mining communities. *Rural Society, 22*(2), 138–152.

Bice, S. (2014). What gives you a social licence? An exploration of the social licence to operate in the Australian mining industry. *Resources, 3*(1), 62–80.

Bice, S. (2015). Bridging corporate social responsibility and social impact assessment. *Impact Assessment and Project Appraisal, 33*(2), 160–166. doi:10.1080/14615517.2014.983710.

Bice, S., & Moffat, K. (2014). Social licence to operate and impact assessment. *Impact Assessment and Project Appraisal, 32*(4), 257–262.

Black, L. (2013). *The social licence to operate: Your management framework for complex times*. Oxford: Do Sustainability.

Blowfield, M. (2005). Corporate social responsibility—The failing discipline and why it matters for international relations. *International Relations, 19*(2), 173–191.

Blowfield, M., & Dolan, C. S. (2008). Stewards of virtue? The ethical dilemma of CSR in African agriculture. *Development and Change, 39*(1), 1–23.

Blowfield, M., & Frynas, J. G. (2005). Setting new agendas: Critical perspectives on corporate social responsibility in the developing world. *International Affairs, 81*(3), 499–513.

Bomsel, O., Börkey, P., Glachant, M., & Lévêque, F. (1996). Is there room for environmental self-regulation in the mining sector? *Resources Policy* 22 (1–2): 79–86. doi:10.1016/S0301-4207(96)00030-X.

Boutilier, R. (2009). *Stakeholder politics: Social capital, sustainable development and the corporation*. Stanford: Stanford University Press.

Boutilier, R. G., & Thomson, I. (2011). *Modelling and measuring the social licence to operate: Fruits of a dialogue between theory and practice*. Accessed May 5, from http://socialicense.com/publications/Modelling and Measuring the SLO.pdf

Bowen, H. R. (1953). *Social responsibilities of the businessman*. New York: Harper & Row.

Braithwaite, J. (2008). *Regulatory capitalism: How it works, ideas for making it work better*. Cheltenham: Edward Elgar.

Branco, M., & Rodrigues, L. (2006). Corporate social responsibility and resourcebased perspectives. *Journal of Business Ethics, 69*(2), 111–132.

Bridge, G., & McManus, P. (2000). Sticks and stones: Environmental narratives and discursive regulation in the forestry and mining sectors. *Antipode, 32*(1), 10–47. doi:10.1111/1467-8330.00118.

Brueckner, M., Durey, A., Mayes, R., & Pforr, C. (Eds.). (2014). *Resource curse or cure? On the sustainability of development in Western Australia*. Heidelberg: Springer.

Brueckner, M., Durey, A., Pforr, C., & Mayes, R. (2014). The civic virtue of developmentalism: On the mining industry's political licence to develop Western Australia. *Impact Assessment and Project Appraisal, 32*(4), 315–326. doi:10.1080/14615517.2014.929784.

Brueckner, M., & Mamun, M. A. (2010). Living downwind from CSR: A community perspective on corporate practice. *Business Ethics: A European Review, 19*(4), 326–348.

Brueckner, M., & Pforr, C. (2011). Western Australia's short-lived 'sustainability revolution'. *Environmental Politics, 20*(4), 585–589.

Brueckner, M., & Ross, D. (2010). *Under corporate skies: A struggle between people, place and profit*. Perth: Fremantle Press.

Chandler, L. (2014). Regulating the resource juggernaut. In M. Brueckner, A. Durey, R. Mayes, & C. Pforr (Eds.), *Resource curse or cure? On the sustainability of development in Western Australia* (pp. 165–177). Heidelberg: Springer.

Chubb, P. (2014). *Power failure. The inside story of climate politics under Rudd and Gillard*. Collingwood, VIC: Black Inc. Agenda.

Chubb, I. (2015). *The science and research priorities and practical research challenges*. Canberra.

Clark, J. M. (1916). The changing basis of economic responsibility. *The Journal of Political Economy, 24*(3), 209–229.

Conservation Council of Western Australia. (2014a, October 14). Industry and investors warned gas fracking will not be welcome in Southwest. *Media Release*. http://ccwa.org.au/media/industry-and-investors-warned-gas-fracking-will-not-be-welcome-southwest - sthash.QHR7wsvI.dpuf

Conservation Council of Western Australia. (2014b, November 27). Mid West local residents tell Norwest at company AGM: Fracking not welcome on the Turquoise Coast. *Media Release*. http://ccwa.org.au/media/mid-west-local-residents-tell-norwest-company-agm-fracking-not-welcome-turquoise-coast?page=2 - sthash.W4JmYIIw.dpuf

Crane, A., & Matten, D. (2007). Corporate social responsibility as a field of scholarship. In A. Crane, & D. Matten (Eds.), *Corporate social repsonsibility* (Vols. I, II and III). London: Sage.

Crane, A., McWilliams, A., Matten, D., Moon, J., & Siegel, D. S. (2008). *The Oxford handbook of corporate social responsibility*. New York: Oxford University Press.

Cutifani, M. (2013, February 4–7). *Mining's contribution to sustainable development*. African Mining Indaba 2013, Cape Town. http://www.icmm.com/document/5043

Dashwood, H. S. (2012). *The rise of global corporate social responsibility: Mining and the spread of global norms*. Cambridge: Cambridge University Press.

Davis, R., & Franks, D. (2014). *Costs of company-community conflict in the extractive sector* (Corporate Social Responsibility Initiative Report No. 66). Cambridge, MA.

Dean, M. (2010). *Governmentality: Power and rule in modern society*. London: Sage Publications.

Detomasi, D. A. (2008). The political roots of corporate social responsibility. *Journal of Business Ethics, 82*(4), 807–819.
Doane, D., & Abasta-Vilaplana, N. (2005). The Myth of CSR. The problem with assuming that companies can do well while also doing good is that markets don't really work that way. *Stanford Social Innovation Review, 3*(3), 23–29.
Douglas, M. (1992). *Risk and blame: Essays in cultural theory*. London: Routledge.
Dunn, J. (2000). *The cunning of unreason: Making sense of politics*. New York: Basic Books.
Eckersley, R. (1998). *Measuring progress: Is life getting better?* Collingwood: CSIRO Publishing.
Franks, D. M., & Cohen, T. (2012). Social licence in design: Constructive technology assessment within a mineral research and development institution. *Technological Forecasting and Social Change, 79*(7), 1229–1240.
Franks, D., Fidler, C., Brereton, D., Vanclay, F., & Clark, P. (2009). *Leading practice strategies for addressing the social impacts of resource developments*. Briefing paper for the Department of Employment, Economic Development and Innovation, Queensland Government. Brisbane.
Friedman, M. (1970, September 13). The social responsibility of business is to increase its profits. *New York Times Magazine*, pp. 122–126.
Gallop, G. (2003). *State Government rejects Ningaloo Reef marina project*. Perth.
Garnaut, R. (2014, May 14). *Power failure*. Speech by Ross Garnaut at the Book Launch of Power Failure, Melbourne.
Gasfield Free Northern Rivers. (2015). *Gasfield free communities*. Accessed March 17, from http://csgfreenorthernrivers.org/csg-free-communities/
Geppert, M., Matten, D., & Walgenbach, P. (2006). Transnational institution building and the multinational corporation: An emerging field of research. *Human Relations, 59*(11), 1451–1465.
Giddens, A. (1990). *The consequences of modernity*. Cambridge: Polity Press in association with Basil Blackwell, Oxford.
Gilding, M., Merlot, E., Leitch, S., Bunton, V., & Glezos, L. (2012). Media framing of the resources super profits tax. *Australian Journal of Communication, 39*(3), 23–40.
Gilles, M., Galloway, D., Horwitz, P., & Larson, A. (2004). Reintegrating coastal development and public health in Western Australia. *EcoHealth, 1*(4), 340–350.
Grabosky, P. N. (1995). Regulation by reward: On the use of incentives as regulatory instruments*. *Law and Policy, 17*(3), 257–282. doi:10.1111/j.1467-9930.1995.tb00150.x.
Grabosky, P. (2013). Beyond responsive regulation: The expanding role of non-state actors in the regulatory process. *Regulation and Governance, 7*(1), 114–123. doi:10.1111/j.1748-5991.2012.01147.x.
Graetz, G., & Franks, D. M. (2015). Conceptualising social risk and business risk associated with private sector development projects. *Journal of Risk Research* (ahead-of-print), 1–21.
Grudnoff, M. (2014). *Fracking the future: Busting industry myths about coal seam gas*. Canberra: Australia Institute.
Guillaume, B., & Charron, S. (1999). *Exploring implicit dimensions underlying risk perception of waste from mining and milling of uranium ores in France*. Fontenay-aux-Roses: Institute for Protection and Nuclear Safety.
Gunningham, N. A., Kagan, R. A., & Thornton, D. (2004). Social license and environment protection: Why businesses go beyond compliance. *Law and Social Inquiry, 29*(2), 307–341.
Habermas, J. (1979). *Legitimation crisis*. London: Heinemann.
Hadji, R., & Sweeney, P. (2014, August 15). Greenough says no fracking. *The Guardian*, p. 4.
Haines, F. (2009). Vanquishing the enemy or civilizing the neighbour? Controlling the risks from hazardous industries. *Social and Legal Studies, 18*(3), 397–415.
Haines, F. (2011). *The paradox of regulation: What regulation can achieve and what it cannot*. Cheltenham: Edward Elgar Publishing.

Hall, N. L. (2014). Can the "social licence to operate" concept enhance engagement and increase acceptance of renewable energy? A case study of wind farms in Australia. *Social Epistemology, 28*(3–4), 219–238. doi:10.1080/02691728.2014.922636.
Hall, N., Ashworth, P., & Devine-Wright, P. (2013). Societal acceptance of wind farms: Analysis of four common themes across Australian case studies. *Energy Policy, 58*(0), 200–208. doi:10.1016/j.enpol.2013.03.009.
Hall, N., Lacey, J., Carr-Cornish, S., & Dowd, A.-M. (2015). Social licence to operate: Understanding how a concept has been translated into practice in energy industries. *Journal of Cleaner Production, 86*(0), 301–310. doi:10.1016/j.jclepro.2014.08.020.
Hamann, R., & Acutt, N. (2003). How should civil society (and the government) respond to 'corporate social responsibility'? A critique of business motivations and the potential for partnerships. *Development Southern Africa, 20*(2), 255–270. doi:10.1080/03768350302956.
Hannam, P., & Nicholls, S. (2015). Planning rule change gives hope to Bulga, other communities facing big mines. *The Sydney Morning Herald.* http://www.smh.com.au/environment/planning-rule-change-gives-hope-to-bulga-other-communities-facing-big-mines-20150706-gi6bvh.html - ixzz3fvjjWhw5
Hart, S. (1997). Beyond greening: Strategies for a sustainable world. *Harvard Business Review, 75*(1), 66–76.
Harvey, B., & Bice, S. (2014). Social impact assessment, social development programmes and social licence to operate: tensions and contradictions in intent and practice in the extractive sector. *Impact Assessment and Project Appraisal, 32*(4), 327–335. doi:10.1080/14615517.2014.950123.
Hasham, N. (2014, October 13). Coal seam gas concerns 'valid', says EPA chief. *The Sydney Morning Herald.* http://www.smh.com.au/nsw/coal-seam-gas-concerns-valid-says-epa-chief-20141013-115913.html - ixzz3fviswJYn
Herman, E. S., & Chomsky, N. (2002). *Manufacturing consent. The political economy of the mass media*. New York: Pantheon Books.
Hoare, R., & Rose, S. (2011). CSG sparks action. *Waste Management and Environment, 22*(5), 38.
Holme, C. (2010). Corporate social responsibility: A strategic issue or a wasteful distraction? *Industrial and Commercial Training, 42*(4), 179–185.
Idemudia, U. (2009). Oil extraction and poverty reduction in the Niger Delta: A critical examination of partnership initiatives. *Journal of Business Ethics, 90*(1), 91–116.
Idemudia, U. (2010). Rethinking the role of corporate social responsibility in the Nigerian oil conflict: The limits of CSR. *Journal of International Development, 22*(7), 833–845.
Igalens, J., & Gond, J. P. (2005). Measuring corporate social performance in France: A critical and empirical analysis of ARESE data. *Journal of Business Ethics, 56*(2), 131–148.
Integrated Risk Governance Council. (2005). *White paper on risk governance. Towards an integrative approach.* Geneva:Integrated Risk Governance Council.
International Council on Mining and Metals. (2003). *10 principles for sustainable development.* Accessed May 21. https://www.icmm.com/our-work/sustainable-development-framework/10-principles
International Standards Organisation. (2009). *Risk management vocabulary.* Geneva: International Standards Organisation.
Jackson, G. (2014). A socio-political perspective on corporate social responsibility: Understanding regulatory substitution and the persistence of irresponsibility. In R. M. Hilty & F. Henning-Bodewig (Eds.), *Corporate Social Responsibility* (pp. 19–31). Berlin: Springer.
Jackson, G., & Apostolakou, A. (2010). Corporate social responsibility in Western Europe: An institutional mirror or substitute? *Journal of Business Ethics, 94*(3), 371–394.
Jacobs, M. (1991). *The green economy: Environment, sustainable development and the politics of the future*. London: Pluto Press.
Jasanoff, S. (1986). *Risk management and political culture. A comparative study of science in the policy context, social research perspectives. Occasional reports on current topics.* New York: Russell Sage Foundation.

Jijelava, D., & Vanclay, F. (2014). Social licence to operate through a gender lens: The challenges of including women's interests in development assistance projects. *Impact Assessment and Project Appraisal, 32*(4), 283–293.

Kaplan, S., & Garrick, B. J. (1981). On the quantitative definition of risk. *Risk analysis, 1*(1), 11–27.

Kellow, A., & Niemeyer, S. (1999). The development of environmental administration in Queensland and Western Australia: Why are they different? *Australian Journal of Political Science, 34*(2), 205–222.

Klein, N. (2007). *The shock doctrine. The rise of disaster capitalism.* London: Allen Lane.

Kotler, P., & Lee, N. (2005). *Corporate social responsibility: Doing the most good for your company and your cause.* Hoboken, NJ: Wiley.

Kuch, D., Ellem, G., Bahnisch, M., & Webb, S. (2013). *Social license and communications report.* Newcastle.

Lacey, J. (2013, February 27). Can you legislate a social licence to operate? *The conversation.* http://theconversation.com/can-you-legislate-asocial-licence-to-operate-10948

Lacey, J., & Lamont, J. (2014). Using social contract to inform social licence to operate: An application in the Australian coal seam gas industry. *Journal of Cleaner Production, 84*(0): 831–839. doi:10.1016/j.jclepro.2013.11.047.

Lacey, J., Moffat, K., Zhang, A., & Ashworth, P. (2014). Earning a social licence for alternative transport fuels. In R. Clark & M. Thomson (Eds.), *Transport fuels from Australia's gas resources* (pp. 207–281). Sydney: UNSW Press.

Lantos, G. P. (2001). The boundaries of strategic corporate social responsibility. *Journal of Consumer Marketing, 18*(7), 595–630.

Lee, M.-D. P. (2008). A review of the theories of corporate social responsibility: Its evolutionary path and the road ahead. *International Journal of Management Reviews, 10*(1), 53–73.

Leyne, L. (2014, June 18). 'Social license' comes home to roost. *Times Colonist.* http://www.timescolonist.com/opinion/columnists/les-leyne-social-licence-comes-home-to-roost-1.1135452

Matten, D., & Crane, A. (2005). Corporate citizenship: Towards an extended theoretical conceptualization. *Academy of Management Review, 30*, 166–179.

Matten, D., & Moon, J. (2008). 'Implicit' and 'explicit' CSR: A conceptual framework for a comparative understanding of corporate social responsibility. *Academy of Management Review, 33*(2), 404–424.

McWilliams, A., & Siegel, D. S. (2001). Corporate social responsibility: A theory of the firm perspective. *Academy of Management Review, 26*(1), 117–127.

Meyer, J., & Rowan, B. (1977). Institutionalized organizations: Formal structure as myth and ceremony. *American Journal of Sociology, 83*(2), 340–363.

Minerals Council of Australia. (2014). *Red tape reduction—A great first step.* Accessed May 21, from http://www.minerals.org.au/news/red_tape_reduction_a_great_first_step

Minerals Council of Australia (MCA). (2013, October 29). Advancing mining and human rights. *Press Release.* http://www.minerals.org.au/news/advancing_mining_and_human_rights

MMSD Project. (2002). *Breaking new ground.* London: Earthscan.

Moffat, K., & Zhang, A. (2014). The paths to social licence to operate: An integrative model explaining community acceptance of mining. *Resources Policy, 39*(0): 61–70.

Morgan, R. K. (2012). Environmental impact assessment: the state of the art. *Impact Assessment and Project Appraisal, 30*(1), 5 14.

Morrison, J. A. (2014). *The social license: How to keep your organization legitimate.* Basingstoke: Palgrave Macmillan.

O'Riordan, L., & Fairbrass, J. (2008). Corporate social responsibility (CSR): Models and theories in stakeholder dialogue. *Journal of Business Ethics, 83*(4), 645–758.

OECD. (2013). *Public consultation on best practice principles for improving regulatory enforcement and inspections: Draft report submitted to the public for comments.* Paris: OECD.

Organ, M. (2014). New tactics see coal seam gas protests gain the upper hand. *The Conversation, 28*, 1–4.
Orltizky, M. (2005). Payoffs to social and environmental performance. *Journal of Investing, 14*(3), 48–52.
Owen, J. R., & Kemp, D. (2013). Social licence and mining: A critical perspective. *Resources Policy, 38*(1), 29–35. doi:10.1016/j.resourpol.2012.06.016.
Panayotou, T. (1994). *Economic instruments for environmental management and sustainable development.* Nairobi: UNEP.
Parker, C. (2002). *The open corporation : Effective self-regulation and democracy.* Cambridge: Cambridge University Press.
Parsons, R., Lacey, J., & Moffat, K. (2014). Maintaining legitimacy of a contested practice: How the minerals industry understands its 'social licence to operate'. *Resources Policy, 41*(0), 83–90. doi:10.1016/j.resourpol.2014.04.002
Parsons, R., & Moffat, K. (2014). Constructing the Meaning of Social Licence. *Social Epistemology, 28*(3–4), 340–363. doi:10.1080/02691728.2014.922645.
Pini, B., Mayes, R., & McDonald, P. (2010). The emotional geography of a mine closure: A study of the Ravensthorpe nickel mine in Western Australia. *Social and Cultural Geography, 11*(6), 559–574.
Porter, M. E., & Kramer, M. R. (2002). The competitive advantage of corporate philanthropy. *Harvard Business Review, 80*(12), 56–69.
Porter, M. E., & Kramer, M. R. (2006). Strategy and society: The link between competitive advantage and corporate social responsibility. HBR Spotlight. *Harvard Business Review, 84*(12), 78–92.
Porter, M. E., & Kramer, M. R. (2007). Strategy and society: The link between competitive advantage and corporate social responsibility—Response. *Harvard Business Review, 85*(6), 136–137.
Presten, J. (2014). If mining conflicts suppress the right of public participation, then can mining be sustainable? *International Journal of Rural Law and Policy, 0*(1), 1–8. http://epress.lib.uts.edu.au/journals/index.php/ijrlp/article/view/3849
Prno, J. (2013). An analysis of factors leading to the establishment of a social licence to operate in the mining industry. *Resource Policy, 38*, 577–590.
Prno, J., & Slocombe, D. S. (2012). Exploring the origins of 'social license to operate' in the mining sector: Perspectives from governance and sustainability theories. *Resources Policy, 37*(3), 346–357.
Rio Tinto. (2013, May 21). *2013 sustainable development. Supporting our licence to operate.* Accessed from http://www.riotinto.com/sustainabledevelopment2013/_pdf/rio_tinto_2013_sustainable_development.pdf
Robertson, J. (2015, May 14). Adani's Carmichael mine: Damage far outweighs benefits, lawyer claims. *The Guardian.* http://www.theguardian.com/australia-news/2015/may/14/adanis-carmichael-mine-damage-far-outweighs-benefits-lawyer-claims
Roche, C., & Mudd, G. (2014). An overview of mining and the environment in Western Australia. In M. Brueckner, A. Durey, R. Mayes, & C. Pforr (Eds.), *Resource curse or cure? On the sustainability of development in Western Australia* (pp. 179–194). Heidelberg: Springer.
Rosa, E. A. (1998). Metatheoretical foundations for post-normal risk. *Journal of Risk Research, 1*(1), 15–44.
Sapolsky, H. M. (1990). The politics of risk. *Daedalus, 119*, 83–96.
Scherer, A. G., & Palazzo, G. (2011). The new political role of business in a globalized world: A review of a new perspective on CSR and its implications for the firm, governance, and democracy. *Journal of Management Studies, 48*(4), 899–931.
Schmidpeter, R. (2013). Corporate social responsibility: A new management paradigm? In J. O. Okpara & S. O. Idowu (Eds.), *Corporate social responsibility* (pp. 171–180). Berlin: Springer.
Schreck, P. (2011). Reviewing the business case for corporate social responsibility: New evidence and analysis. *Journal of Business Ethics, 103*(2), 167–188.
Shamir, R. (2010). Capitalism, governance, and authority: The case of corporate social responsibility. *Annual Review of Law and Social Science, 6*(1), 531–553.

Slovic, P., Fischhoff, B., & Lichtenstein, S. (1984). Behavioral decision theory perspectives on risk and safety. *Acta Psychologica, 56*(1), 183–203.
Solomon, F., Katz, E., & Lovel, R. (2008). Social dimensions of mining: Research, policy and practice challenges for the minerals industry in Australia. *Resources Policy, 33*(3), 142–149. doi:10.1016/j.resourpol.2008.01.005.
Stigler, G. J. (1971). The theory of economic regulation. *The Bell Journal of Economics and Management Science, 2*(1), 3–21. doi:10.2307/3003160.
Taylor, L. (2015, January 14). Federal court asked to overturn Adani mine approval due to impact on Great Barrier Reef. *The Guardian*. http://www.theguardian.com/environment/2015/jan/14/federal-court-asked-to-overturn-adani-mine-approval-impact-great-barrier-reef
Tempel, A., & Walgenbach, P. (2007). Global standardization of organizational forms and management practices? What new institutionalism and the business systems approach can learn from each other. *Journal of Management Studies, 44*(1), 1–24.
Thomson, I., & Boutilier, R. G. (2011). The social licence to operate. In P. Darling (Ed.), *SME mining engineering handbook* (pp. 1779–1796). Littleton: Society for Mining, Metallurgy and Exploration.
Toft, K. H. (2015). Liberal CSR and new Marxist criticism. In S. O. Idowu, C. S. Frederiksen, A. Y. Mermod, & M. E. J. Nielsen (Eds.), *Corporate social responsibility and governance* (pp. 303–316). Cham: Springer International Publishing.
Towie, N. (2013, September 13). Green light for Roe Hwy angers conservation groups. *WAtoday.com.au*. http://www.watoday.com.au/wa-news/green-light-for-roe-hwy-angers-conservation-groups-20130913-2tpmy.html - ixzz3fvhfdwWN
Turkina, N., Neville, B., & Bice, S. (2015). Rediscovering divergence in developing countries' CSR. In D. Jamali, C. Karam, & M. Blowfield (Eds.), *CSR in developing countries: A development-oriented approach*. Sheffield: Greenleaf.
United Nations. (2000). *The global compact*. New York: United Nations.
Vial, V. (2007). Corporate social responsibility: A third way. *Global Business and Organizational Excellence, 26*(6), 18–27.
Vogel, D. (2008). Private global business regulation. *Annual Review of Political Science, 11*(1), 261–282.
Walker, C. (2012). Coal, gas and coal seam gas. *Chain Reaction, 115*, 12–15.
Wesley, A. (2014). *The socio-political construction and experience of corporate social responsibility (CSR): An investigation into the conflict surrounding the James Price Point LNG precinct, Kimberley, Western Australia*. PhD thesis. Perth: School of Management, Curtin University.
Wesley, A., & MacCallum, D. (2014). The political economy of corporate social responsibility in the resource sector in Western Australia. A case study of the proposed James Price Point LNG precinct. In M. Brueckner, A. Durey, R. Mayes, & C. Pforr (Eds.), *Resource curse or cure? On the sustainability of development in Western Australia* (pp. 59–73). Heidelberg: Springer.
Wilburn, K., & Wilburn, R. (2011). Achieving social license to operate using stakeholder theory. *Journal of Business Ethics, 4*, 3–16.
Williams, P., Gill, A., & Ponsford, I. (2007). Corporate social responsibility at tourism destinations: Toward a social license to operate. *Tourism Review International, 2*, 133–144.
Willis. (2013). *Mining market review 2013*. London: Willis.
Yakovleva, N. (2005). *Corporate social responsbility in the mining industries*. Cornwall: MPG Books.
Yim, M., & Vaganov, P. A. (2003). Effects of education on nuclear risk perception and attitude: Theory. *Progress in Nuclear Energy, 42*(2), 221–235.

Part II
CSR and Sustainability

Sustainable Logistics: A Framework for Green Logistics and City Logistics

Carsten Deckert

1 Introduction

The global flow of materials and goods is steadily increasing. While the yearly production of goods increased by the factor 5.3 from 1960 to 2008, the export of goods in real terms increased by the factor 15.6 during the same period (Bpb 2014). This disproportionate increase of exports is mainly caused by the increasing global interlacing of economies in the course of globalization and the liberalization of world trade. Additionally offshoring activities of transnational corporations relocating facilities around the globe and transporting semifinished products between them contribute to the increase in the flow of goods.

In an increasingly globalized world logistical activities gain more and more importance. A study by Roland Berger Strategy Consultants predicts that the market for logistics will grow 2.4–3.0% annually from 2014 to 2019 (Doll et al. 2014). At the same time the negative effects of logistical activities on the environment and society are bound to grow. One example is the emission of greenhouse gases: The World Economic Forum estimates that logistical activities already cause 2800 mega-tonnes of greenhouse gases annually and predicts that the freight transport carbon emissions will grow over the coming years (WEF 2009). At the same time the trend of urbanization with more people living in cities than in rural areas (UN 2014) makes deliveries to end-customers more and more difficult. These developments demand an approach for sustainable logistics.

The term sustainability is usually attributed to the forestry official Carl von Carlowitz who employed it to the field of forestry in the eighteenth century. In this field it means a balance between a shortterm harvest and a longterm use of the resource of wood. Thus, the origin of the term is tied to the consumption of

C. Deckert (✉)
Innovation and Production Management, Hochschule Düsseldorf (HSD), Düsseldorf, Germany
e-mail: carsten.deckert@hs-duesseldorf.de

H. Lu et al. (eds.), *Building New Bridges Between Business and Society*, CSR, Sustainability, Ethics & Governance, DOI 10.1007/978-3-319-63561-3_4

resources (Pufé 2012). In the twentieth century the report of the Brundtland Commission (WCED 1988) emphasizes the longterm orientation of sustainability when it defines a sustainable development as "a development that meets the needs of the present without compromising the ability of future generations to meet their own needs" and broadens the concept to also include emissions when it states that an "unsustainable development may arise not only from overuse of certain commodities but from manufactured goods that are potentially polluting". Elkington (1999) in his Triple Bottom Line (TBL)-approach adds the dimensions of environmental protection and social equity to the dimension of economic prosperity for a company.

In logistics, efforts in sustainability usually focus on environmental sustainability. Environmental sustainability can be distinguished into "strong" and "weak" sustainability. "Strong" sustainability opts for sufficience (limited consumption) and subsistence (self-supply) (compare e.g. Paech 2012). "Weak" sustainability usually works with the principles of resource efficiency (more value for less resource input) and eco-efficiency (more value with less environmental impact). Ideally economic growth should be decoupled from resource consumption and environmental impacts (Pufé 2012; Kanning 2013). A third approach is the cradle-to-cradle-principle by Braungart and McDonough (2009) respectively the principle of eco-effectiveness. In this approach products should be designed in such a way that the used resources either go back into the technical cycle as secondary raw materials or into the biological cycle as biological nutrients. The approaches of "weak" sustainability and cradle-to-cradle both rely on innovation and new technologies to improve eco-efficiency and eco-effectiveness.

Approaches in logistics usually fall under the category of "weak" sustainability or cradle-to-cradle. That means that they try to increase resource efficiency, eco-efficiency and eco-effectiveness. This is in line with the Brundtland-report which explicitly states that the motto for industry is "Producing More with Less" (WCED 1988).

In the sphere of logistics the discussion about sustainability usually gravitates around the two terms Green Logistics and City Logistics. The following article will describe the approaches currently falling under those categories drawing heavily on the experiences made during a German book project conducted at the end of 2014/the beginning of 2015 (Deckert 2015). Firstly the impacts of logistics on the environment in the form of resource consumption and emissions are explained. After that a framework for green logistics and city logistics is introduced based on the logistical functions transportation, warehouse management and packaging using amongst others examples from the book project. Finally pressing problems of sustainable logistics are discussed from which possible research directions are drafted.

2 Impacts of Logistics on the Environment

Logistics has negative impacts on the environment concerning resource consumption and emissions of harmful substances (see Fig. 1). Resources which are consumed can be energy, raw materials, water, air and space. Emissions of logistical

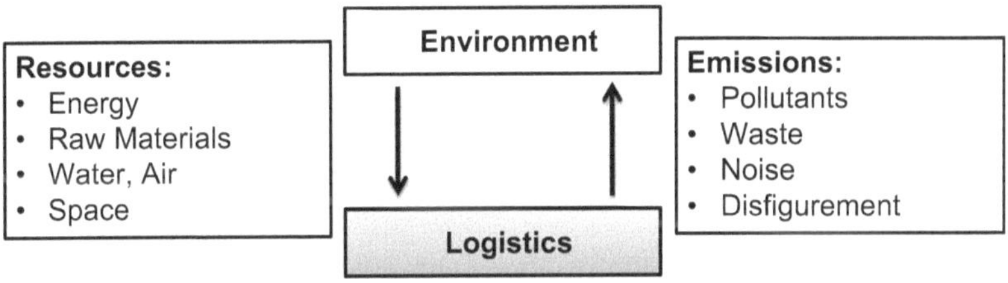

Fig. 1 Impacts of logistics on the environment (in accordance with Willke 2006)

activities can appear as pollutants (e.g. exhaust fumes or greenhouse gases), waste, noise and disfigurement of the landscape (in accordance with Willke 2006).

Resource consumption and emissions can lead to so called negative external effects, i.e. the costs of the resource consumption respectively the emissions are not fully covered by the causal agent. They have to be carried by other parties, often society as a whole. These environmental costs are more and more internalized, i.e. charged to the causal agents by political instruments. Political instruments to internalize environmental costs can be taxes and fees, standards, certificates or sanctions, such as decrees and bans (Hartwig 2007; Willke 2006).

The incentive for companies to deal with environmental protection is steadily increasing. Schwister (2010, own translation) gives the following definition for environmental protection: "Environmental protection aims at the preservation of the livelihoods, mainly those which will have to undergo an overly alteration without human intervention". This definition is in line with the concept of sustainability as defined in the introduction. In this paper environmental protection is divided into resource consumption and environmental friendliness. Schwister (2010, own translation) also distinguishes these two fields calling them "environmental consumption of quantitative livelihoods" and "environmental pollution of qualitative livelihoods".

2.1 Consumption of Resources

Resources can be defined as substances or things which have a benefit for society. So what one considers as a resource depends largely on the function of the substance or thing and can vary over time. Resources can be divided into non-renewable or stock resources and renewable or flow resources. For non renewable resources there is "no possibility of their being replenished on a timescale of relevance to human society" while renewable resources are "naturally renewed within a sufficiently short time-span to be of use for society" (Daniels et al. 2012).

The non-renewable resources can be divided into those which are consumed during use (e.g. oil, coal or natural gas) and those which are theoretically and/or practically recyclable (e.g. metallic minerals). Concerning the renewable resources it is useful to distinguish between resources which are not depleted during use and thus are non-critical (e.g. solar energy, tides, wind and waves) and resources for which there is a danger of overuse and which can therefore be considered critical (e.g. fish and other wild animals, forests). The renewability of these resources depends on the intensity of use and the human management efforts (Daniels et al. 2012).

The original idea of sustainability refers to the critical renewable resources. Here a permanent and stable use pattern which allows the resource to be renewed is preferable. This can be difficult for freely available resources (such as fish) since there we often encounter the "Tragedy of the commons"-problem. For non-critical resources sustainability is by definition a given. For non-renewable resources sustainability can be achieved in the short to medium term if the rate of new discovery of resource reserves is as big as or bigger than the rate of consumption (Schwister 2010). This period can be prolonged by a more efficient use of resources, the increase of the recycling rate of raw materials (if feasible) or the changeover to renewable resources to be in accordance with the principles of sustainability. For critical renewable resources overuse again has to be avoided.

The improvement of the consumption of resources can be termed resource efficiency (see Fig. 2). Resource efficiency describes the benefit of a product or service in terms of its resource consumption and contains the following three components (Deckert and Fröhlich 2014):

- *Raw material efficiency* aims at reducing the raw material input for a given output and usually decreases material cost. It also includes all measures for reuse and recycling of resources.
- *Energy efficiency* aims at the reduction of the energy input for a given performance and usually decreases energy cost.
- *Environmental stewardship* aims at an economical and considerate exploitation of natural resources in general, especially resources with no or only low prices (e.g. water, air and land), and also includes the increased use of renewable materials and energies while simultaneously avoiding their overuse.

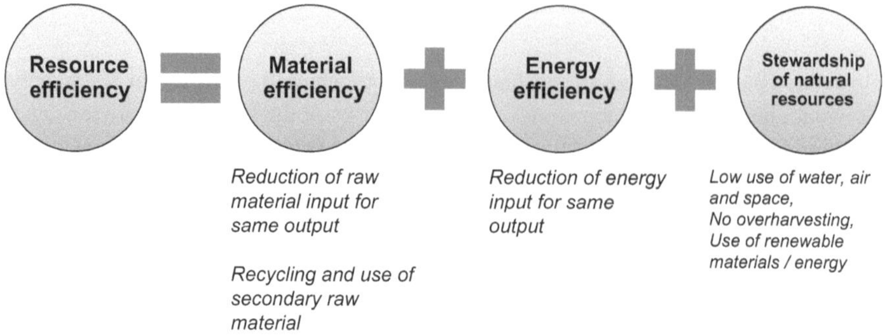

Fig. 2 Components of resource efficiency

Material and energy efficiency have a positive impact on the running costs through a reduction of material and energy costs. But they can require investments into new technologies. Environmental stewardship does not necessarily lead to decreased costs. But it could lead to cost reductions in case of increasing prices or political measures such as fees and can secure the longterm availability of a resource. This usually requires an adjustment of existing technologies or investments into new technologies.

2.2 Environmental Friendliness Concerning Emissions

In general emissions can lead to environmental damages in the following spheres (vgl. z.B. McNeill 2001; Schwister 2010; Townsend et al. 2008):

- *Air (Atmosphere)*
  Emissions into the atmosphere can either lead to local pollution of the air with direct consequences (e.g. exhaust fumes such as carbon monoxide or particulate matter) or global impacts with indirect consequences (e.g. climate change through greenhouse gases such as carbon dioxide or depletion of the ozone layer through chlorofluorocarbons).
- *Water (Hydrosphere)*
  Environmental damage in the hydrosphere comprises e.g. contamination through biological and chemical wastes, acidification, eutrophication (excess nutrients), oxygen deficiency and siltation. In certain areas of the world too much use of water can lead to desiccation of bodies of water or depression of ground water.
- *Ground/Soil (Lithosphere and Pedosphere)*
  The ground can be spoiled e.g. by soil contamination (e.g. through toxic chemicals), soil erosion and baring (e.g. through deforestation) or soil sealing (e.g. through road construction).
- *Flora and Fauna (Biosphere)*
  Environmental damage in the biosphere can lead to the destruction of natural habitats and the reduction of biodiversity. Transportation activities can lead to bioinvasions (introductions of alien species into an ecosystem). As a result of these damages species can become endangered or go extinct.

The transfer of the sustainability concept from the sphere of resources to the field of emissions is problematic. The strict demand of sustainability would mean that it wouldn't be allowed to have more emissions during a certain period than can be degraded in that period including all reaction products (see Schwister 2010). This seems to be difficult in some fields (e.g. concerning carbon dioxide emissions) and impossible in others (e.g. land use for infrastructure). For this reason many measures today aim for an improvement of eco-efficiency which means that one gets the same value with less environmental impact concerning emissions.

3 Green Logistics Framework

In the literature slightly different definitions of logistics exist. The Council of Supply Chain Management Professionals (CSCMP 2013) e.g. describes the term in its glossary as follows: "The process of planning, implementing, and controlling procedures for the efficient and effective transportation and storage of goods including services, and related information from the point of origin to the point of consumption for the purpose of conforming to customer requirements. This definition includes inbound, outbound, internal, and external movements." Jahns and Schüffler (2008) give an overview over 13 German definitions most of which have in common that logistics deals with the material flow and the corresponding information flow, has planning, realization and controlling as its central activities and comprises certain logistical functions.

In the literature on logistics typically the following logistical functions are distinguished (see e.g. Pfohl 2010; Schulte 2009):

- *Transportation* is the movement of goods between different locations using different modes of transport. Modes of freight transport can be road transport, rail transport, pipeline transport, inland waterway transport, ocean or sea shipping, air transport or a combination of these (intermodal transport) (Chopra and Meindl 2013; Grant 2012).
- *Warehouse management* respectively warehousing can be defined as "that part of a company's logistics system that stores products (raw materials, parts, goods in progress, finished goods) at certain nodes in the firm's supply chain, and provides information to the firm on the status, condition and disposition of items being stored" (Grant 2012). Warehousing which deals with the physical aspects of storage can be distinguished from inventory management which deals with all the decisions exclusively concerning inventory of goods.
- *Packaging* can be divided into the functional use of packaging systems and the creation of loading units using loading devices such as pallets or containers. Functions of packaging are protection and preservation, storage and transportation facilitation, market appeal, identification and information, containment and user convenience (ten Hompel et al. 2007; Lee and Lye 2003).

Green logistics aims at improving the resource consumption and the environmental friendliness of logistics. While Heiserich et al. (2011, p. 11) see the target of green logistics in the responsible way of dealing with resources, Dekker et al. (2012) set the focus more on emissions when they define green logistics as the "study of practices that aim to reduce the environmental externalities, mainly related to greenhouse gas emissions, noise and accidents, of logistics operations". Many other authors, e.g. Koch (2012), McKinnon (2012) and Wittenbrink (2014), include both targets—resource consumption and environmental friendliness—into their definitions.

According to the author of this paper green logistics is a logistical concept which mainly deals with the environmental dimension of the Triple Bottom Line and aims

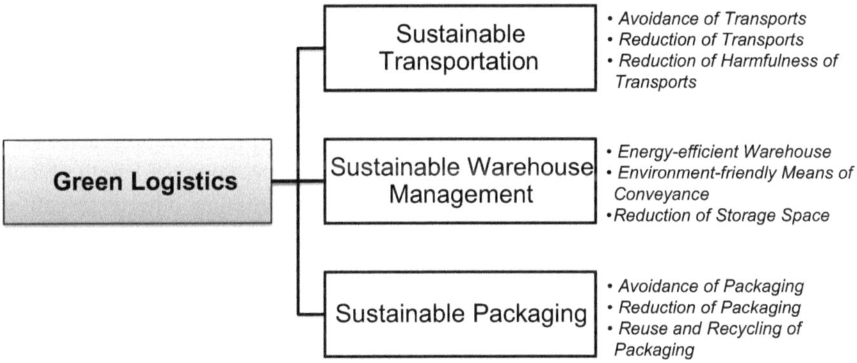

Fig. 3 Green logistics-framework

at the alignment of the logistical functions with the targets of improvement of resource efficiency and reduction of emissions and related environmental damage. The task of green logistics is the fulfillment of the "6 R" of logistics under consideration of resource consumption and environmental friendliness, i.e. to make the right quantity of the right product available at the right place and the right time in the right condition for the right cost while keeping resource consumption and emissions as low as possible.

In the following sub-chapters the impacts of the different logistical functions on the environment are described, the addressed core issues are highlighted and examples of possible countermeasures are given (see Fig. 3).

3.1 *Sustainable Transportation*

The core issues of sustainable transportation are fuel consumption and emissions into the air, i.e. exhaust fumes and greenhouse gas emissions. In addition there are noise emissions. The necessary infrastructure leads to land use and impacts on flora and fauna (Koch 2012), e.g. soil sealing through roads or destruction of natural habitats through the construction of canals. Further emissions into air, water or ground are possible in case of accidents. In case of global supply chains the problem of bio-invasions can arise as animals travel "as stowaway" alongside the freight or ballast water in ocean vessels. This contribution mainly focusses on the effects of the failure-free use of different modes of transportation and neglects impacts of infrastructures and accidents. In political decisions on transport infrastructure as well as in risk analyses of transportation these, of course, have to be additionally taken into consideration.

According to the World Economic Forum transport activities cause 90% of the 2800 mega-tonnes of greenhouse gas emissions caused by logistical activities. Road freight alone causes 57% of the logistical greenhouse gas emissions. But based on transport performance measured in tkm air freight causes the highest amount of the emissions followed by road, rail and ships (WEF 2009).

According to McKinnon (2012) the main parameters which affect the environmental friendliness for transportation are modal split, average handling factor (as a measure for the number of changes of transportation mode), average length of haul, average load on laden trips, average percentage of empty runs, other externalities per vehicle-km and per unit of throughput (e.g. noise, accidents), energy efficiency (of transportation vehicle) and emissions per unit of energy.

The environmental friendliness of transportation can thus be increased by the following approaches:

1. Avoidance of transports
2. Reduction of transports (number of transports and length of transport route)
3. Reduction of harmfulness of transports

The avoidance of transports can be achieved by digitalization and remote transmission of products which has already been achieved for products such as software, music, books and movies. Through the increased use of 3D-printers this trend might also spread to other product categories albeit the effects might be limited. The sharing of goods might also contribute to the avoidance of transports.

The reduction of transports can be achieved by the reduction of the number of transports and the reduction of the length of routes. The reduction of the number of transports can be achieved by bundling volumes (e.g. through milk runs), by matching goods flows and reducing empty runs and by increasing the capacity utilization, e.g. through optimized use of stowage space (see Bode et al. 2011; Bretzke and Barkawi 2012).

The reduction of the length of routes can be achieved on a strategic level by the design of the supply chain network. Possible approaches are e.g. relocation of offshored activities and nearshoring, optimized warehousing levels and number of warehouses as well as different distribution of production activities to facilities. On a more operative level approaches can be improved by transport routing and use of intelligent transport system (ITS) to track and trace deliveries and vehicle fleets (see Bode et al. 2011).

Environment-friendly means of transport and modal split can lead to a reduced harmfulness of transports. Examples for measures are environment-friendly propulsion technologies (e.g. electromobility), optimized combustion engines, optimized aerodynamic resistance of vehicles, training of drivers and higher automation of vehicles (DHL 2010; Rees 2014; Stodick 2015; Kampker et al. 2015). An environment-friendly modal split can be achieved by switching to environment-friendlier means of transport such as rail or vessels (e.g. short sea shipment in coastal areas).

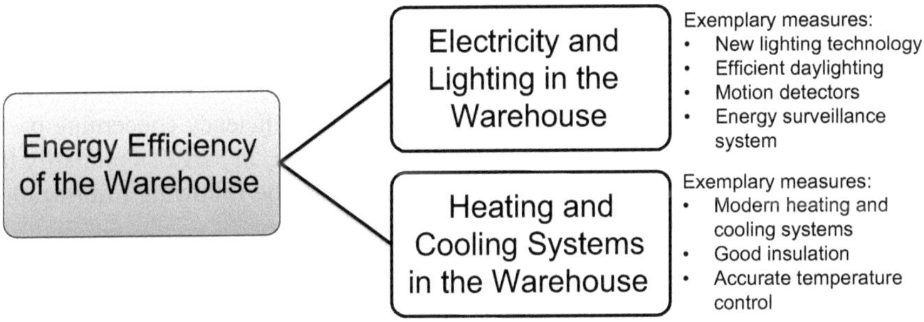

Fig. 4 Energy efficiency in the warehouse (in accordance with DHL 2010)

3.2 Sustainable Warehouse Management

The focus of sustainable warehouse management is on energy efficiency and land use. About 10% of the 2800 mega-tonnes of greenhouse gas emissions of logistical activities are caused by logistical buildings according to the estimates of the World Economic Forum (WEF 2009). Starting points for sustainable warehouse management are environment-friendly storage, environment-friendly conveyance and reduction of storage space.

Environment-friendly storage aims at a high energy efficiency of the warehouse (see Fig. 4). This can be achieved by measures in the following fields (DHL 2010):

- Electricity and lighting in the warehouse (e.g. periodic cleaning of lamps, energy-efficient lighting technology, efficient use of daylight, movement-sensitive lighting, energy monitoring systems)
- Heating and cooling systems (e.g. modern heating and cooling systems, good insulation, accurate temperature control)

In addition to these measures the means of conveyance can be switched to more environment-friendly propulsion technologies (e.g. electrical fork lifters).

These measures all have positive impacts not only on the sustainability of a warehouse, but also on the cost of warehouse usage (reduction of cost for lighting, heating and cooling). However they increase the cost of warehouse construction (higher cost for buildings and devices). In the construction of warehouses these calculations should already be incorporated as life cycle cost. Additionally the enhanced use of renewable resources as construction material can be included. E.g. the company Schachinger in Austria built a high rack warehouse made from wood combining the criteria construction ecology, life cycle cost, energy efficiency and workplace quality (Hiebl 2015).

3.3 Sustainable Packaging

From 2005 to 2011 the total quantity of packaging waste of the EU-27 rose from 78.6 to 79.9 million tonnes. This means that on average every citizen of the EU-27 generated 159 kg of packaging waste in 2011. The shares of packaging per weight are 40% paper and cardboard, 20% glass, 19% plastics, 15% wood and 6% metals (Eurostat 2013).

The focus of sustainable packaging is on material efficiency concerning packaging design and on waste treatment after packaging use. In logistics primarily the transport and outer packaging are addressed. Usually the following prioritization for sustainable packaging is given:

1. Prevention of packaging
2. Reduction of packaging quantity and harmfulness
3. Reuse through reusable packaging systems
4. Recycling and recovery of packaging material

Through an adequate packaging design packaging material can be prevented or reduced. Potentially hazardous materials can be substituted by less hazardous materials. Moreover non-renewable materials can be exchanged for renewable materials. The main task of sustainable packaging design is to balance the new function of sustainability with the established packaging functions. Negative impacts on protection and preservation (e.g. lower tensile strength through lower quantity of packaging material), on storage and transport facilitation (e.g. more effort in stapling and handling), on market appeal (e.g. smaller packages suggest less content) as well as on convenience of use (e.g. lower stability of packaging during usage) have to be avoided (Schweig 2015).

Reusable transport packaging usually has to be returned and prepared for reuse. This means that reusable packaging does not only have advantages, such as e.g. reduction of packaging cost in the long run and adherence to legislation, but can also have negative consequences, such as e.g. higher production cost and capital commitment as well as costs for returning, repair, cleaning and administration (Heiserich et al. 2011).

Recycling usually means regaining the used raw materials and reintroducing them into the production process as secondary raw material. For this the product form has to be dissolved in contrast to reuse of packaging. Recycling is mainly a question of collecting and sorting waste since usually only single-variety waste material can be recycled and a question of energy-intensity of the recycling process. So for some materials thermal recovery in waste incineration plants is preferable at the current technological level.

4 City-Logistics

The term city logistics can be defined in different ways. McKinnon (2012) classifies city logistics under the term green logistics. A narrow definition usually only includes the use of urban consolidation centers (UCCs) at the edge of town to consolidate the flow of goods into the city (Jahns and Schüffler 2008; Piontek 2009; Schulte 2009). Zsifkovits (2013) adds the use of telematics and informatics to this narrow definition. According to Crainic et al. (2009) the bundled transport into cities should also take place in environment-friendly vehicles.

A wide definition of city logistics is given by Taniguchi et al. (2001): "City logistics is the process for totally optimising the logistics and transport activities by private companies in urban areas while considering the traffic environment, traffic congestion and energy savings within the framework of a market economy." It is obvious though that city logistics focusses on transport management in urban areas. The target of city logistics thus is according to Crainic et al. (2009) the reduction of the "nuisances associated to freight transportation in urban areas while supporting their economic and social development".

The term city is not defined in a consistent way in the international literature (UN 2014). But a few characteristics of cities are usually a certain size (e.g. number of inhabitants), density and concentration, heterogenity and differentiation as well as permanence (see Bathelt and Glückler 2003; Daniels et al. 2012; Leser 1997; Seto et al. 2013; UN 2014). There is a global trend of urbanization which led to more people living in cities than in rural areas since 2007. The United Nations forecast that until 2050 about two thirds of the global population will live in cities (UN 2014). Furthermore there is a trend for ever bigger cities or so called mega-cities with more than ten million inhabitants, especially in Asia (Gebhardt et al. 2011).

With increasing size cities have increasing problems in supplying goods and disposing of wastes. For logistics in urban areas there are currently the problems of smaller deliveries and just-in-time-deliveries (JIT), inefficiencies of deliveries (e.g. at the ramp of the recipient), overlapping of passenger and delivery traffic mainly on the streets, traffic congestion and higher emissions of noise, exhaust fumes and particulate matter, delivery restrictions in cities and proliferation of the municipal area (BVL 2014; Schulte 2009; DHL 2010).

City logistics can be defined as the improvement of the supply of urban areas with goods (incl. return logistics), especially by using existing public infrastructure. Targets of city logistics are the relief of the strain on the transport infrastructure (e.g. less traffic congestion, less accidents) and the reduction of (direct) emissions (e.g. noise, particulate matter). So its focus is on existing transportation infrastructure and can be distinguished from city planning in this way.

City logistics currently uses the following effects and measures to achieve its targets (see Fig. 5):

- **Bundling of transports and matching of the flows of goods** can be achieved by freight consolidation in urban consolidation centers (UCCs) at the edge of town which usually involves the cooperation and coordination of different carriers and the use of intelligent transport systems (ITS) to optimize transport routes and track and trace deliveries (Crainic et al. 2009; DHL 2014; Piontek 2009; Schulte 2009; Zsifkovits 2013).
- **Rectification in Space and Time** can be achieved on the one hand by installing separate transport lanes or using means of transport in the air or underground such as e.g. drones or CargoCap (Handelsblatt 2014; Macho and Menn 2014; Stein et al. 2014). On the other hand deliveries can be made outside of the rush

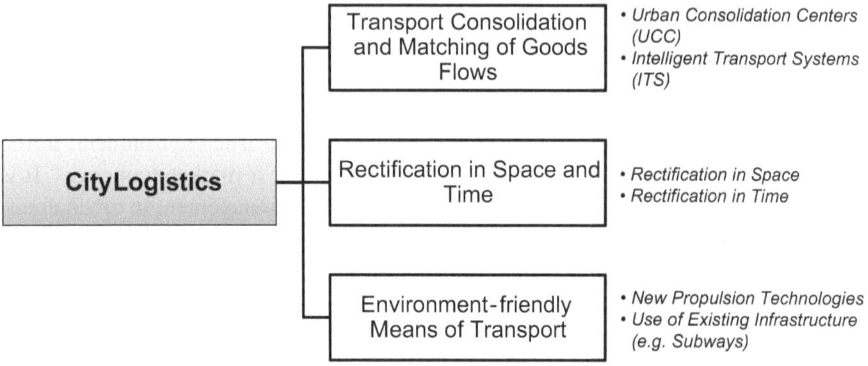

Fig. 5 Effects and measures of city-logistics

hour (e.g. nighttime deliveries) or can be placed in PUDO (pickup and drop-off) solutions (DHL 2010, 2014).

- **Use of Environment-friendly Means of Transport** includes the development and application of environment-friendly propulsion technologies such as electromobility, natural gas or cargo bikes (see Kampker et al. 2015; Rees 2014; Stodick 2015) or the use of existing passenger infrastructure such as subway or tram for freight transport (DHL 2014), e.g. CarGoTram for the supply of goods to the VW facility in Dresden (Roth 2012).

5 Problem Discussion and Possible Research Directions

In Fig. 6 the main impacts of the logistical functions transportation, warehousing and packaging on resource consumption and emissions are summarized. The examples in this article show that there are already some solutions, but major problems still remain in the following three areas:

- Solving the trade-offs between logistical functions and integrating the neglected social perspective
- Examining the underlying issues
- Developing an approach for implementation and measurement

5.1 Solving the Trade-offs Between Logistical Functions and Integrating the Social Perspective

The described examples from the different logistical functions show that companies already started to address several of the problems in green logistics and city logistics. Nevertheless these examples are often isolated measures taking into account

Logistical Functions	Resources	Emissions
Transportation*	• Fuel Consumption	• Exhaust Fumes • Greenhouse Gases • Noise • Bio-Invasions (Ocean Transport)
Warehousing	• Energy Consumption • Warehouse Space	• Greenhouse Gases (Scope 2) • Disfigurement of Sites
Packaging	• Material Consumption • Energy Consumption	• Waste

** Only well-functioning operation without consideration of malfunctioning or accidents and impact of infrastructure*

Fig. 6 Main logistical impacts on resource consumption and emissions

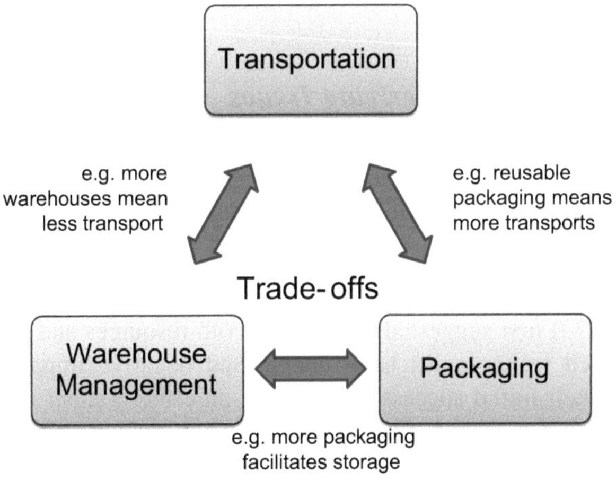

Fig. 7 Trade-offs of sustainable logistics

only one or a limited set of targets. In the future trade-offs between the functions of transportation, warehousing and packaging have to be put into focus (see Fig. 7). Three examples covering the three possible areas of trade-offs should indicate possible research directions:

1. In supply chain network design there is a minimum of greenhouse gas emissions which results from the number of warehouses and transports. More warehouse locations lead to higher greenhouse gas emissions from warehousing, but at the

same time decrease the transportation activities leading to less greenhouse gas emissions from transport (DHL 2014).
2. As described above packaging design can lead to the prevention of packaging, but at the same time runs the risk of lowering the storage and transportation facilitation.
3. The 4flow Supply Chain Management Study shows that the decision about the use of reusable packaging systems can only be made under consideration of the type of good to be transported and the length of transport distances (Gross et al. 2013; Zesch 2015).

As the term green logistics implies the environmental impacts are at the core while social issues of logistics are often neglected. Social issues of logistics are e.g. noise emissions from transportation activities and their impact on human well-being, accidents of delivery vehicles and safety issues impacting other traffic participants and leading to severe damage to people and property as well as work conditions and remuneration of logistics workers. Future research should integrate these issues into a holistic and integrated model of sustainable logistics. Research in this field should lead to a deeper understanding of the mechanisms and interdependencies of measures for sustainable logistics in the different functions and the TBL dimensions.

5.2 *Examining the Underlying Issues*

The main root causes of the grave impacts of logistics and other economic activities on the environment which require action are according to the author of this article as follows:

- There are *limits to growth* as the report of the Club of Rome (Meadows and Meadows 1972) first suggested concerning both resources and emissions.
- The employed technologies have unexpected negative impacts which often result from their initial success in adoption, an effect Tenner (1997) described as the *revenge of unintended consequences of technology*.
- Environmental goods are often public goods affected by negative external effects, a phenomenon described by Hardin (1968) as the *tragedy of the commons*.

These causes are accelerated by the three main drivers namely demographic growth, economic growth and increasing energy consumption and are considered as the tough underlying issues by the author to be tackled. At the moment many approaches in sustainable logistics just cover the field where sustainability and cost reduction overlap. Simultaneously some sustainable logistics measures fail due to shortterm cost reasons. Research in this area should lead to a deeper understanding of underlying patterns of environmental and social problems and reveal better ways to design incentives by managers as well as politicians.

5.3 *Developing an Approach for Implementation and Measurement*

As the examples described above show there is already an awareness of sustainability in logistics. An international survey of Bundesvereinigung Logistik (BVL) e.V., the biggest logistical association in Germany, among 1757 supply chain executives shows that for 55% of the respondents sustainability is already a component of the logistics strategy, but that many supply chain executives still struggle with implementing it into daily activities (Handfield et al. 2013). From the examples gathered a three-step approach for implementation can be suggested:

- *Main issues*: Identifying the main issues to be addressed can take the main logistical issues into account (see Fig. 6), but needs to be adapted to the specific situation of the company. E.g. UPS not only focusses on energy consumption, emissions and road safety concerning their transport activities, but also on water consumption because their fleet of appr. 96,000 vehicles needs to be washed (Stodick 2015). Moreover a stakeholder analysis can support the evaluation by increasing transparency about all the impacted persons and organisations.
- *Metrics*: The precise and reliable measurement of environmental impacts is a main prerequisite for the design of effective measures. In transportation and warehousing e.g. the measurement of greenhouse gas emissions can lead to measures for improvement, such as the development and use of three-temperature-zone delivery trucks to decrease greenhouse gas emissions by the wholesaler Lekkerland (Effertz 2015). Measurements of greenhouse gas emissions in transportation should be based on DIN EN 16258 and a well-to-wheel-evaluation taking all direct emissions (scope 1), indirect emissions (scope 2) and emissions of subcontractors (scope 3) into consideration and transforming them in CO_2-equivalents (Schmied and Knörr 2013). Resource consumption can be analyzed using life cycle assessment and substance flow maps.
- *Measures*: Based on the metrics for the main issues effective measures can be designed that aim at the desired outcomes and can be improved in a step-by-step way using the defined metrics.

This research should lead to procedures to implement sustainable logistics measures and methods to measure and compare the results.

Pagell and Shevchenko (2014) demand that the artificial division of Supply Chain Management and Sustainable Supply Chain Management—and the author would like to add between Logistics and Sustainable Logistics—must be overcome to leave the "Focus of the Familiar" behind and to take the trade-offs and underlying issues of sustainability into consideration. According to EN ISO 9000 (CEN 2005) the quality of a product or service can be defined as the "degree to which a set of inherent characteristics fulfils requirements". The author's vision of sustainability is that it will be integrated into this definition of quality as one of the requirements to be fulfilled for high quality logistics.

References

Bathelt, H., & Glückler, J. (2003). *Wirtschaftsgeographie* (2. Aufl ed.). Stuttgart: UTB.

Bode, W., Ziegler, A., Hermsmeier, J., & Hocke-Anbeh, A. (2011). *Praxisleitfaden "Grüne Logistik"*, Osnabrück.

Bpb (Bundeszentrale für politische Bildung). (2014). *Entwicklung des grenzüberschreitenden Warenhandels*. Accessed December 17, 2014, from http://www.bpb.de/nachschlagen/zahlen-und-fakten/globalisierung/52543/entwicklung-des-warenhandels

Braungart, M., & McDonough, W. (2009). *Cradle to cradle: Re-making the way we make things*. New York: Vintage Books.

Bretzke, W.-R., & Barkawi, K. (2012). *Nachhaltige Logistik – Antworten auf eine globale Herausforderung*. Berlin: Springer.

BVL. (2014). *Grünbuch Nachhaltige Logistik in urbanen Räumen*. Wien: Eigendruck.

CEN (Comité Européen de Normalisation). (2005). *Quality management systems – Fundamentals and vocabulary (ISO 9000:2005)*. Brussels: CEN.

Chopra, S., & Meindl, P. (2013). *Supply chain management: Strategy, planning, and operations*. Harlow: Pearson.

Crainic, T. G., Ricciardi, N., & Storchi, G. (2009). Models for evaluating and planning city logistics systems. *Transportation Science, 43*(4), 432–454.

CSCMP (Council of Supply Chain Management Professionals). (2013). *Supply chain management terms and glossary*. Accessed December 17, 2014, from http://cscmp.org/sites/default/files/user_uploads/resources/downloads/glossary-2013.pdf

Daniels, P., Bradshaw, M., Shaw, D., & Sidaway, J. (2012). *An introduction to human geography* (4. Aufl. ed.). Harlow: Pearson.

Deckert, C. (2015). *CSR & Logistik. Spannungsfelder Green Logistics und City Logistik*. Berlin: Springer.

Deckert, C., & Fröhlich, E. (2014). Green logistics: Framework zur Steigerung der logistischen Nachhaltigkeit. *Supply Chain Management II*, 13–18.

Dekker, R., Bloemhof, J., & Mallidis, I. (2012). Operations research for green logistics – An overview of aspects, issues, contributions and challenges. *European Journal of Operational Research, 219*, 671–679. doi:10.1016/j.ejor.2011.11.010.

DHL. (2010). *Delivering Tomorrow – Zukunftstrend Nachhaltige Logistik*. Bonn: Eigendruck.

DHL. (2014). *Logistics trend radar. Delivering insight today – Creating value tomorrow*. Bonn: Eigendruck.

Doll, A., Friebel, D., Rückriegel, M., & Schwarzmüller, C. (2014). *Global logistics market*. München: Roland Berger Strategy Consultants.

Effertz, L. (2015). Nachhaltige Transportlogistik bei Lekkerland. In C. Deckert (Ed.), *CSR & Logistik. Spannungsfelder Green Logistics und City Logistik*. Berlin: Springer.

Elkington, J. (1999). *Cannibals with forks. Triple bottom line of 21st century business*. North Mankato: Capstone.

Eurostat. (2013). *Packaging waste statistics*. Accessed June 23, 2015, from http://ec.europa.eu/eurostat/statistics-explained/index.php/Packaging_waste_statistics

Gebhardt, H., Glaser, R., Radtke, U., & Reuber, P. (2011). *Geographie. Physische Geographie und Humangeographie* (2. Aufl.). Heidelberg: Spektrum.

Grant, D. B. (2012). *Logistics management*. Harlow: Pearson.

Gross, W., Zesch, F., Gelau, T., Hayden, C., Bötel, M., & Brock, M. (2013). *Costs and benefits of green logistics – 4flow supply chain management study 2013*. Berlin: 4flow.

Handelsblatt. (2014, September 30). Post liegt im Drohnen-Wettlauf vorn. In *Handelsblatt News am Abend* (p. 4).

Handfield, R., Straube, F., Pfohl, H.-C., & Wieland, A. (2013). *Trends and strategies in logistics and supply chain management – Embracing global logistics complexity to drive market advantage*. Bremen: DVV Media Group.

Hardin, G. (1968). Tragedy of the commones. *Science, 162*, 1243–1248.

Hartwig, K.-H. (2007). Umweltökonomie. In D. Bender et al. (Hrsg.), *Vahlens Kompendium der Wirtschaftstheorie und Wirtschaftspolitik* (7. Auflage, pp. 129–170). München: Vahlen.
Heiserich, O.-E., Helbig, K., & Ullmann, W. (2011). *Logistik. Eine praxisnahe Einführung* (4. Aufl.). Berlin: Springer.
Hiebl, I. (2015). Leuchtturm LT1 – Europäisches Leuchtturm-Projekt im nachhaltigen Hallenbau. In C. Deckert (Ed.), *CSR & Logistik. Spannungsfelder Green Logistics und City Logistik*. Berlin: Springer.
Jahns, C., & Schüffler, C. (2008). *Logistik – Von der Seidenstraße bis heute*. Wiesbaden: Wissensreihe des Siegfried-Vögele-Instituts.
Kampker, A., Deutskens, C., Maue, A., & Hollah, A. (2015). Elektromobile Logistik am Beispiel StreetScooter. In C. Deckert (Ed.), *CSR & Logistik. Spannungsfelder Green Logistics und City Logistik*. Berlin: Springer.
Kanning, H. (2013). Nachhaltige Entwicklung – Die gesellschaftliche Herausforderung für das 21. Jahrhundert. In A. Baumast & J. Pape (Hrsg.), *Betriebliches Nachhaltigkeitsmanagement* (pp. 21–43). Stuttgart: Eugen Ulmer.
Koch, S. (2012). *Logistik – Eine Einführung in Ökonomie und Nachhaltigkeit*. Berlin: Springer.
Lee, S., & Lye, S. (2003). Design for manual packaging. *International Journal of Physical Distribution & Logistics Management, 33*(2), 163–189.
Leser, H. (1997). *Diercke Wörterbuch Allgemeine Geographie* (12. Aufl ed.). München: Dtv.
Macho, A., & Menn, A. (2014). Hilfreiche Überflieger. *Wirtschaftswoche, 40*(29.9.2014), 78–82.
McKinnon, A. (2012). Environmental sustainability: A new priority for logistics managers. In A. McKinnon, A. Browne, & A. Whiteing (Eds.), *Green logistics: Improving the environmental sustainability of logistics* (2nd ed.). London: Kogan Page.
McNeill, J. R. (2001). *Something new under the sun: An environmental history of the world in the 20th century*. London: Penguin Books.
Meadows, D., & Meadows, D. (1972). *The limits to growth. A report for the club of Rome's project on the predicament of mankind*. New York: Universe Pub.
Paech, N. (2012). *Befreiung vom Überfluss. Auf dem Weg in die Postwachstumsökonomie*. München: Oekom Verlag.
Pagell, M., & Shevchenko, A. (2014). Why research in sustainable supply chain management should have no future. *Journal of Supply Chain Management, 50*(1), 44–55.
Pfohl, H.-C. (2010). *Logistiksysteme – Betriebswirtschaftliche Grundlagen*. Berlin: Springer.
Piontek, J. (2009). *Bausteine des Logistikmanagements* (3. Aufl.). Herne: NWB.
Pufé, I. (2012). *Nachhaltigkeit*. Konstanz: UTB.
Rees, J. (2014). Kampf dem Kollaps. *Wirtschaftswoche, 39*, 72–77.
Roth, K. (2012). *Über Schienen ins VW-Werk. Forum – Das Wochenmagazin*. Accessed December 17, 2014, from http://archiv.magazin-forum.de/uber-schienen-ins-vw-werk/
Schmied, M., & Knörr, W. (2013). *Berechnung von Treibhausgasemissionen in Spedition und Logistik gemäß DIN EN 16258. Begriffe, Methoden, Beispiele* (2. Aufl). Bonn: DSLV.
Schulte, C. (2009). *Logistik. Wege zur Optimierung der Supply Chain* (5. Aufl.). München: Vahlen.
Schweig, C. E. (2015). Nachhaltige Verpackungsentwicklung. In C. Deckert (Ed.), *CSR & Logistik. Spannungsfelder Green Logistics und City Logistik*. Berlin: Springer.
Schwister, K. (2010). *Taschenbuch der Umwelttechnik* (2. Aufl.). Leipzig: Hanser.
Seto, K. C., Parnell, S., & Elmqvist, T. (2013). A global outlook on urbanization. In T. Elmqvist et al. (Eds.), *Urbanization, biodiversity and ecosystem services. Challenges and opportunities: A global assessment*. Berlin: Springer. doi:10.1007/978-94-007-7088-1_1.
Stein, D., Beckmann, D., Siefer, T., & Stein, R. (2014). *Automatischer Güterverkehr im Untergrund. Ein verkehrstechnisches Zukunftsszenario* (Working Paper der EUFH Brühl).
Stodick, K. (2015). Reduzieren, Vermeiden und Ausgleichen – Nachhaltigkeitsmanagement bei UPS. In C. Deckert (Ed.), *CSR & Logistik. Spannungsfelder Green Logistics und City Logistik*. Berlin: Springer.

Taniguchi, E., Thompson, R. G., Yamada, T., & van Duin, R. (2001). *City logistics. Network modelling and intelligent transport systems*. Bingley: Emerald.
ten Hompel, M., Schmidt, T., & Nagel, L. (2007). *Materialflusssysteme. Förder- und Lagertechnik* (3. Aufl.). Berlin: Springer.
Tenner, E. (1997). *Why things bite back: Technology and the revenge of unintended consequences*. New York: Vintage Books.
Townsend, C. R., Begon, M., & Harper, J. L. (2008). *Essential of ecoclogy* (3rd ed.). Oxford: Blackwell.
UN (United Nations, Department of Economic and Social Affairs, Population Division). (2014). *World urbanization prospects: The 2014 revision, highlights (ST/ESA/SER.A/352)*. New York: UN.
WCED (World Commission on Environment and Development). (1988). *Our common future*. Oxford: Oxford University Press.
WEF (World Economic Forum). (2009). *Supply chain decarbonization. The role of logistics and transport in reducing supply chain carbon emissions*. Accessed June 23, 2015, from http://www.weforum.org/pdf/ip/SupplyChainDecarbonization.pdf
Willke, G. (2006). *Pocket Wirtschaft. Ökonomische Grundbegriffe* (2. Aufl.). Bonn: bpb.
Wittenbrink, P. (2014). *Transportmanagement. Kostenoptimierung, Green Logistics und Herausforderungen an der Schnittstelle Rampe* (2. Aufl.). Berlin: Springer.
Zesch, F. (2015). Einweg geht mehr Weg – eine ökologische und ökonomische Analyse zum Einsatz von Einweg- und Mehrwegbehältern unter Berücksichtigung von Transportentfernungen. In C. Deckert (Ed.), *CSR & Logistik. Spannungsfelder Green Logistics und City Logistik*. Berlin: Springer.
Zsifkovits, H. E. (2013). *Logistik*. Konstanz: UTB.

Sustainable Assortment Policy: Possibilities of Differentiation and Profiling for the Food Sector

Christoph Willers and Victoria Aydin

1 Introduction

The food retail industry in Germany as the most important sales channel for food achieved an overall turnover of approx. 175 billion euros (only food) in 2012 (TradeDimensions 2013). The sector is marked by few great actors. The four key players (Edeka, REWE, Schwarz Gruppe, Aldi) unite a sales portion of about 85% (Federal Cartel Authority 2014, p. 10) and, moreover, realise the determining sales increases in the market. This structure leads to an intensive competition amongst the traders, as well as to a relatively low prize level for food and therefore to a pressure on the margins of the food industry. Due to the tension, trading companies are looking for possibilities of differentiation which will have a positive effect on the results.

Beside the current trends in the food sector, like "convenience", "health, fitness and wellness", and "premium and pleasure", the subjects "sustainability, naturalness and authenticity" gain in importance. The "value-revival" of authenticity and origin is originally the base for the development of the organic sector. Meanwhile this frame is increasingly complemented or even partially replaced by the subjects of sustainability or Corporate Social Responsibility (CSR), not least by the expectations of different stakeholders for taking social and ecological responsibility on the part of enterprises (cf. Willers 2015, p. 225).

With respect to the variety of different foods from all over the world, sustainability offers basically a helpful orientation which guarantees (subjective) security for the consumer. Furthermore the clear conscience after purchasing in a sustained manner is an important incentive to buy. Therefore the propensity to sustainable consumer behaviour becomes an increasingly strong differentiating characteristic.

C. Willers (✉) • V. Aydin
CBS Cologne Business School, Köln, Germany
e-mail: c.willers@cbs.de; victoria.aydin@hotmail.com

H. Lu et al. (eds.), *Building New Bridges Between Business and Society*, CSR, Sustainability, Ethics & Governance, DOI 10.1007/978-3-319-63561-3_5

In this way consumers satisfy a huge number of affective demands, i.e. the emotional component of a food. Concerning the food consumption, studies show that in particular demands on value, specific feature, trust, superiority and health dominate (cf. Willers 2007, p. 243).

Therefore the social and ecological sustainability management of an enterprise plays an increasingly important role concerning the purchase decision of the consumer (cf. Webb and Mohr 1999, p. 230 following). For about 80% of the German consumers social and ecological criteria for the product choice are "very important" or "quite important". Sustainability has become a competitive advantage. As a decisive criterion for a product purchase, sustainability is ranked second closely behind quality, but still ahead of the price aspect (cf. Unterbusch 2011, p. 211).

Enterprises of German food trade react to this change of demands with various measures. Full-range provider like REWE and Edeka appear on the market with sustainability labels, regional products as well as with employer image campaigns. Examples might be on the one hand the sustainability weeks of REWE Group which are supposed to strengthen the consciousness of the consumers for environment-friendly and socially acceptable consumption (cf. REWE 2014a) or on the other hand the strategic partnership for comprehensive environment protection of Edeka and WWF whose aim is the reduction of the ecological footprint of Edeka (cf. EDEKA 2014).

For discounters it seems more difficult to reconcile sustainability and low prices. In addition, the negative press in Germany concerning the discussions about the employer's behaviour still continues to have an effect (cf. Willmann 2013, p. 36). Nevertheless, the suppliers of the discounter Aldi are committed to compliance with "social standards in the production" and Aldi sells products with the label of the navy Stewardship Council (MSC) as well as Fairtrade products (cf. ALDI 2014).

In view of the mentioned subjects in the CSR context, central trends for the food retail industry can be identified. These have a direct influence on the strategic orientation of the subject sustainability and also consequences for the assortment politics of food retailers.

2 The Development of Sustainable Products

2.1 Omnipresence of the Subject

Sustainability in the food value chain is in a positive sense increasingly becoming mainstream. The discussion focuses on keywords like Carbon Foot Print, Animal Welfare, Biodiversity, Green Logistic, Employees' rights or Corporate Citizenship. Nevertheless, the term "sustainability" or "sustainable food consumption" is still a very vague concept. At times it is used inflationary and might seem lacking in content (cf. Willers and Weber 2011, p. 22).

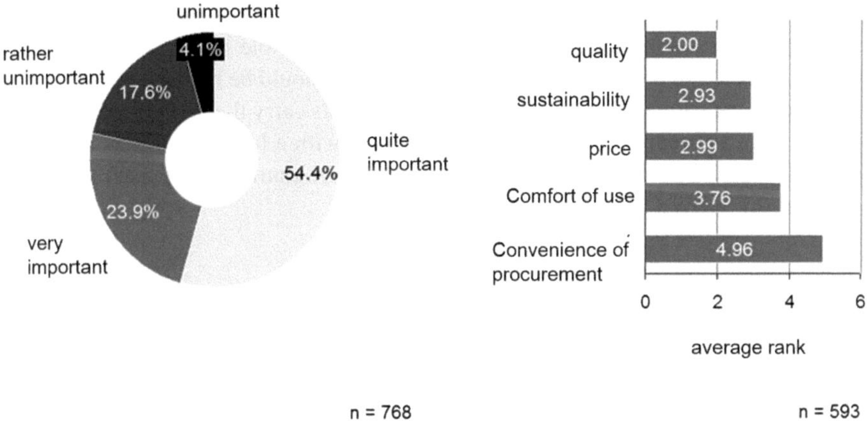

Fig. 1 Relevance of sustainability from consumer view. Source: Unterbusch (2011, p. 211)

Reisch's comprehensive definition reads: "(...) sustainable food consumption as a choice for food which is beneficial and life enhancing for individuals, society and the planet" (2010, p. 1). However, this comprehensive perspective has rarely been realised in practice so far (cf. Meyer-Höfer and Spiller 2013, p. 1).

Although sustainability is not the first decision criterion for consumers (see Fig. 1), the subject functions as a "hygiene factor": "Through the purchase of sustainable products one can make at least one small contribution for a better world" (Willmann 2013, p. 35, translation mine). The public demands that companies show a "minimum of sustainable corporate governance" (ibid., p. 36). Not only sustainability as a subject, but also the embodied or at least propagated lifestyle in large parts of the society "of being sustainable" has apparently become "normal". However, normal and taken-for-granted offers or behaviour patterns face the problem of selective perception. Something special, fascination or—generally expressed—the value of being new is missing (Willers 2015, p. 226).

Therefore the subject itself is no longer suitable to be used as a differentiation criterion in the food economy. Meanwhile the actors are facing the challenge to meet the minimum requirements without any problems and at the same time the need to differentiate and extend one's competitive edge through far-reaching measures. REWE Group can be seen as the pioneer in German food trade concerning differentiation efforts for sustainable subjects. In particular the sales channel REWE supermarket has developed a sustainable commitment in recent years. Their CSR measures include among others the opening of CO_2-neutral supermarkets as well as social activities (cf. REWE 2014a).

The introduction of the sustainability label PRO PLANET serves as a direct differentiation at the point-of-sale: "The new label of REWE marks products whose production, processing or use places far less burden on environment and society. The aim is to promote sustainable consumption and to offer products with sustainable value at attractive prices. Through the purchase of PRO PLANET you can make a contribution to the wellbeing of humans and environment" (REWE 2014b).

By means of a "hot spot-analysis" (repetition every 3 years) relevant ecological and social problems along the supply chain or along the whole life cycle of a product can be analysed in order to identify which problem should be reduced or eliminated by suitable measures. Meanwhile about 500 products carry the PRO PLANET label (status as of February 2014). Primarily, products with a large number of customer contacts (e.g., fruit and vegetables, eggs, coffee, yoghurt, toilet paper) are being selected (cf. REWE 2014c).

2.2 *Low Willingness to Pay*

German grocers apparently are not afraid to market sustainable products in an offensive manner. The risk is too high that, according to the prize-sensitive food market in Germany, these more expensive products can lead to customers' buying reticence. Although investigations show a positive attitude pattern concerning the relevance of sustainability for consumers (Fig. 1), the market is not yet driven by supply, i.e. the availability of sustainable produced products does not necessarily mean that there is a demand for these goods (attitude-behaviour-bias).

Survey results are often interpreted in such a way as that acceptance is equated with purchase or in negative case with non-purchase. However, a differentiation between a "social acceptance" and a "willingness to buy" is neglected. Nevertheless, in case of non-occurrence of the "willingness to buy", the economic sustainability of an enterprise can be considerably impaired (Willers 2015, p. 230).

Empiric studies show "that attitudes are sometimes good, sometimes weak and sometimes generally unsuitable predictive factors for actual (buying) behaviour" (Wiswede 2000, p. 314). In particular studies concerning the willingness to pay can be seen in that context, because "an interview in a hypothetical situation can often lead to an exaggeration of the stated willingness to pay which does not correspond to the true willingness to pay" (Sterzing 2014, p. 16). Cowe and William call this the "30:3 syndrome". Although 30% of the customers have a positive attitude towards the purchase, the market share is only at 3% (2000, p. 5). The results lead to a more complex handling of smaller loads on the distribution side. Additionally, there is the risk of high transcripts for traders, particularly for fresh and ultra fresh food. Inevitably the question arises of how sustainable produced products should be perceived in the market while these currently rather represent an "undercover offer" in the food market.

2.3 *Significance of Supply Chain*

The already identified hesitant attitude concerning the implementing of CSR-subjects within the product assortment is also reflected within the individual functional areas. Fröhlich points to the fact that especially the area of supply chain management faces completely new challenges and searches for suitable concepts and measures which "fix and implement this point of view in the external and internal supply chain" (2011, p. 9). Different studies conclude that e.g. the implementing of CSR-measures in the supply chain not only reduces procurement costs, but also has a positive impact on the performance of a company (cf. Lee 2010). In addition, there is an interrelation between an economically viable action of every single member of the supply chain and its value for the society (cf. Linton et al. 2007, p. 1079).

Carter and Rogers define sustainable supply chain as "the strategic, transparent integration and achievement of an organization's social, environmental, and economic goal in the systematic coordination of key interorganizational business process for improving the long-term economic performance of the individual company and its supply chains" (2008, p. 368). To ensure a successful implementation of sustainable commitment along the supply chain, it is necessary to adapt the structure of objectives and strategies as well as to consider an appropriate risk management (cf. ibid., p. 369). In the process, the "creation of win-win situations between the different supply chain partners (...) is often overshadowed by the high investment volume which must be provided by the management for sustainable commitment" (Fröhlich 2011, p. 12, translation mine). Anyhow it is pointed to the fact that a social, ecological and economic engagement leads to a relatively high balance (Carter and Rogers 2008, p. 371).

Along with the cost side, the supply chain gains in importance in particular in the context of a possible loss of control for trading ventures. The availability of products (quantitatively/qualitatively) becomes more difficult, e.g. in the area of certificated and sustainably produced raw materials like cocoa or palm oil. Due to the increasing globalization of the supply chains, the risk of less visibility, control and transparency exists for the food-value-chains. Akerlof (1970) describes sustainably produced food as "trust goods". The sustainable production of goods is sensually imperceptible for non-experts. The "recognition" for consumers is currently only possible through suitable labelling. Therefore, trading ventures need to view and control the production process (incl. the "hot-spots") along the entire value chain. Traceability along the value chain becomes increasingly important concerning the quality assurance and additionally as a "guarantee signal" of sustainable operations (Willers 2015, p. 232).

For the implementation of CSR enterprises "have come to include ecological as well as social aspects in the supplier's evaluation" (Fröhlich 2011, p. 16, translation mine). In addition to the general fulfilment of legal requirements, a "return on risk management" can be achievable as soon as this approach serves as a differentiation criterion and competitive advantages arise. Especially an optimised risk management system and with it not only the assurance of the required delivery capability, but also above all the compliance with CSR-specific requirements stress the great economic value added to trade of this approach.

In this context vertical integrations receive an increasing relevance. Verticalisation or vertical integration basically circumscribes a process in which enterprises integrate upstream or downstream stages of production or sales which make a value-added contribution towards the product competence. In the food economy—in particular in German trade—the trend towards a stronger vertical integration of stages of the value chain is clearly recognisable in recent years and seem to be further increasing.

2.4 Regionality as the "New Sustainability"

The bio-sector is faced with a break. Not only conventional campaigners with sustainably produced goods move closer to the "bio-pioneers" according to the perception of the consumers, but also sustainability is in the meantime well on the way to have an even higher level of attention than bio. Although in Germany food with bio-label must contain 95% ingredients of ecological cultivation, the EC-Organic Production Regulation does not consider social criteria within the production process (cf. BMEL 2014). However, the consumer expectations have developed in such way that not only ecological products, but also a sustainable and ethically "correct" product offers are taken for granted.

Nevertheless, bio products are not just competing with sustainable products, but at the same time the openness for regional offers of conventional production increases. If claims are sufficiently satisfied through various disciplines, the interest in bio decreases. In a way proximity beats bio—both in terms of products and geographical. Additionally the regionality approach competes with sustainability. While regionality has an identity in the perception of the public—compact, handy and near—this is exactly the problem of sustainability. Sustainability can rather be described by associations like nonspecific, broad and unfamiliar.

According to the comparison (Table. 1) two strategical options (besides the option of the total renunciation) arise for German food trade: Either a concentration and thus a prioritisation occurs in an area. Referring to this, the prime aim is to profit from the advantages especially for a differentiation as well as to take measures to minimize the disadvantages. Or one offers a combined approach, while key issues are linked in content, e.g. in terms of an integration of sustainability and regionality or "bio from the region".

2.5 Increasing Stakeholder Demands

The German food economy—and therefore also the food trade—is certainly a prime example for how a branch can be kept under media observation. Especially traders as the interface with consumers, and with it as the distributers of food, have to take responsibility. In German trade this leads to visible changes. After a long period during which the German trade was not characterized by its open and proactive communication, a new way of thinking on the part of the merchants has now established itself. Increasing demands on the wide range of in-depth

Table 1 Comparison sustainability, regionality and bio in the food sector

	Sustainability	Regionality	Bio
Advantages	• Omnipresence of the subject • Integrated approach • Connecting factor for cross-cutting activities in enterprises (not limited to assortment policy)	• High (basic) level of trust in regional products • Possibility of differentiation for regional traders • Short delivery distances as well as strengthening the regional economy	• High (basic) level of trust in bio products • Widespread awareness of the bio-label • Societal permeation with further sales potential
Disadvantages	• Unclear definition and content • High variety of labels • Limited product range	• Problem of geographical limits • Regional raw material availability • Focus on vegetable and animal products and low-processed goods	• Focus on ecological aspects • Need for imports due to lacking domestic supply • Decreasing added value due to high sales volumes in the market

Source: Willers (2015, p. 234)

information of the stakeholders lead to a new communicative understanding on the trade side. Stakeholders are any internal and external groups with specific claims and which are "(...) indirectly or directly affected by enterprise activities currently or in the future" (Heinrich and Schmidpeter 2013, p. 13, translation mine).

A core thesis of the strategic management is that the stakeholders with their demands represent one of the entrepreneurial starting points. If resistance occurs to a product offer and to the brand portfolio, one must look for new ways. Therefore the stakeholders with their demands are always the central focus of strategic decisions. When demands of one or several stakeholders change, they must be analysed concerning the effects on the enterprise. On this basis a suitable action plan is to be determined in order to constantly meet the requirements.

Trading ventures must realise which stakeholders function as "drivers" of sustainability and in turn as addressees of information. Basically the group of consumers represents the most important stakeholder group, because they finally decide on the level of demand. In addition, the groups of the media and NGOs (Non Governmental Organizations) are considered to be particularly significant due to their public and media effect.

In the action field of information and communication the necessary transparency and credibility towards the stakeholders are focused: "The consumer finally decides at the point-of-sale which measures seem really sustainable and which are only lip service in his perception" (Willmann 2013, p. 35, translation mine). Three of four consumers are willing "to change the vendor in favour of a guaranteed and extensive offer of goods from sustainable production" (Unterbusch 2011, p. 217, translation mine).

In order to avoid an information overload with the communicants, the news selection need to be restricted to the most important facts for the target group. While a low level of activation is not sufficient to initiate cognitive processes, an intermediate level combined with a positive emotionalisation provide the basis for an efficient processing of information. If the level of the stimulus further rises, cognitive processes in turn are negatively influenced (cf. Wiswede 2000, p. 306). The PRO PLANET label of REWE takes exactly this principle of conciseness within the process of perception into account by the reduction of communication contents to central facts like "protecting biodiversity" or "water-saving production".

The more credible a source of information is or is perceived, the better the communicant will accept the arguments (cf. Hribal 1999, p. 203). Thommen (1990, p. 132) sees credibility as a central guiding principle of entrepreneurial activities. Only by a maximum transparency an essential basis for the credibility can be built which in turn is the foundation for trust ("trust chain"). In practice there is often a correlation between a missing credibility and the willingness to communication. In this connection, safety signals, e.g. in the context of a tracking management, are relevant trust anchors.

On account of the existing trust characteristics of sustainable food, independent labels can help consumers to make an informed purchase decision. "Simple advertising statements of the suppliers or self-created quality marks in contrast are difficult to measure for consumers concerning the credibility." (Meyer-Höfer and Spiller 2013, p. 7, translation mine). Although labelling is of central importance for the distribution of sustainable products, a study shows, "that there are only few labels which have become accepted in the market and have obtained consumer confidence" (ibid., translation mine). The drastically risen number of environment-, bio- and quality-labels rather led to excessive demand of the consumers. More than 80% of the consumers consider the number of sustainability labels as too high (cf. IFH and IBH 2010).

3 Impact on the Assortment Policy

Sustainability management as a conceptual action for the solution of economical, ecological and social challenges (so-called "three-column approach") will be only successful if it integrates into the core business of a trading venture and becomes the business case. Isolated sustainability programmes or activities can give impulses for the assortment policy, but do not lead to lasting—and with it sustainable—solutions. Trading ventures are caught between "duty" and "choice".

The observations have demonstrated that the integration of sustainable products within the assortment policy in the food trade is still handled with reluctance. This currently dominating approach can be characterised as a "duty approach". For trading ventures operating in this context, sustainability is seen as "a must". Traders pick up this restated demand construct opportunistically within the scope of corporate goals. In addition, the breadth and depth of the product ranges confirms that it is a niche within the assortment policy. This is also shown in the goods presentation and in the attention for the articles at the point-of-sale. Often the results are average shelve place profitabilities. Sustainable production is confronted with significant higher unit costs in the market due to the used technology. Hamatschek (2013, p. 24) speaks of "righttech" versus "hightech". Traders who are inconsistent and pursue a self-forced hybrid form will not be successful in the long term, because they cannot take advantage of economies of scale.

The objective of this "duty approach" is the fulfilment of minimum requirements on consumer side. On this occasion, accusations of "greenwashing" are often the result. Therefore PR methods are criticised, because they want to give the enterprise an ecologically friendly and responsible image in public, very much in the spirit "more appearance than substance" (cf. Laufer 2003). In order to prevent the impression of greenwashing, sustainability must form the core and the basic principle of corporate action. Only this allows a credible communication of a suitable corporate philosophy which goes beyond the striving for maximum profit.

More detailed is the stronger linkage and optimisation of sustainably produced products with the aspects of the assortment policy ("choice"). This ensures that sustainability is more than just the performance of one's duty. Existing subjects along the value added chain can be used within this integrated management approach for a value-scooping approach. Considering the background of the earlier described competitive pressure in the German food trade, a suitable sustainability offer is used as a differentiation criterion and with it as an USP. This can only be successful if the value "sustainability" is incorporated into the assortment philosophy and strategy. It is then essential to convey it internally through measures. These might be e.g. defined standards for the purchase department, which proportion of sustainable products within an article group need to be achieved. In addition, supplier's evaluations are also suitable to check with reference to specific key indicators whether actors of the supply chain could be moved for improvements. This can in turn create incentives for employees. Internally these kind of measures positively support the "willingness" of the employees.

In turn externally a recognisable added value must be communicated for the customers. The challenge for the (product) management consists in convincing the stakeholders of the credibility of the social-ecological commitment. The brand management of the retail brand faces "the challenge to create a consistent brand picture which credibly integrates the aspect sustainability without causing resistance." (Unterbusch 2011, p. 218, translation mine). Therefore on one hand clear achievement proofs must be provided and in turn these need to be communicated transparently. On the other hand confirmed and objective certificates, labels, rankings or tests ensure the credibility and therefore the effect of the mentioned strategy.

4 Conclusion

Consumers expect enterprises to fulfil their responsibility to guarantee and constantly improve the sustainability achievements of their products. In order to be successful, enterprises must be able to understand and control their ecological, social and economical (non-) performance. Against the backdrop of the previous executions, hardly any company can afford not to include sustainability in its future strategic orientation. The challenge on commercial side consists in putting food on the market whose social-ecological production is clearly proved and which consumers can trust—and therefore the food need to be produced in a responsible and sustainable manner. Nevertheless, the own economic sustainability must be guaranteed in all activities.

Anybody who includes the value "sustainability" in its corporate or product brand, is able to gain a real competitive advantage. Therefore trading ventures should consider the indicated trends in the complex theme of sustainable products to take the opportunity and at the same time reduce potential risks. Individual measures are no longer sufficient to communicate the competences concerning sustainability—rather innovative solutions as well as credible and suitable overall concepts are required.

References

Akerlof, G. A. (1970). The market for "lemons": Quality uncertainty and the market mechanism. *The Quarterly Journal of Economics, 84*(3), 488–500.

ALDI. (2014). *Wir handeln mit Verantwortung*. Retrieved June 16, 2015, from https://unternehmen.aldi-sued.de/de/verantwortung

BMEL. (2014). *Bio-Siegel*. Retrieved June 22, 2014, from http://www.bmel.de/DE/Landwirtschaft/Nachhaltige-Landnutzung/Oekolandbau/_Texte/Bio-Siegel.html

Carter, C., & Rogers, D. (2008). A framework of sustainable supply chain management: Moving toward new theory. *International Journal of Physical Distribution and Logistics Management, 38*(5), 360–387.

Cowe, R., & Williams, S. (2000). *Who are the ethical consumers*. Manchester: Co-operative Bank.

EDEKA. (2014). *EDEKA und WWF sind Partner für Nachhaltigkeit*. Retrieved June 16, 2015, from https://www.edeka.de/EDEKA/de/edeka_zentrale/verantwortung/nachhaltigkeit/wwf/edeka_nachhaltigkeit.jsp?

Federal Cartel Authority. (2014). *Sektoruntersuchung "Nachfragemacht im Lebensmitteleinzelhandel"*. Retrieved October 6, 2014, from http://www.bundeskartellamt.de/Sektoruntersuchung_LEH.html;jsessionid=B5D6A15FB99A446FB1C3739EC86062D2.1_cid371?nn=4592442

Fröhlich, E. (2011). Nachhaltigkeit und Corporate Social Responsibility in der Supply Chain – Eine Einführung. In E. Fröhlich, T. Weber, & C. Willers (Eds.), *Nachhaltigkeit in der unternehmerischen Supply Chain* (pp. 9–21). Cologne: Fördergesellschaft Produktmarketing.

Hamatschek, J. (2013). *Lebensmittelmanagement*. Stuttgart: Ulmer.

Heinrich, P., & Schmidpeter, R. (2013). Wirkungsvolle CSR-Kommunikation – Grundlagen. In P. Heinrich (Ed.), *CSR und Kommunikation* (pp. 1–25). Berlin: Springer Gabler.

Hribal, L. (1999). *Public-Relations-Kultur und Risikokommunikation: Organisationskommunikation als Schadensbegrenzung*. Forschungsfeld Kommunikation (Bd. 19). Konstanz.

IFH Institut für Handelsforschung, & IBH Retail Consultants. (2010). *Nachhaltigkeit im Handel*. Cologne.
Laufer, W. S. (2003). Social accountability and corporate greenwashing. *Journal of Business Ethics, 43*(3), 253–261.
Lee, H. (2010). Don't tweak your supply chain – Rethink it end to end. *Harvard Business Review*, 1–8.
Linton, J., Klassen, R., & Jayaraman, V. (2007). Sustainable supply chains: An introduction. *Journal of Operation Management, 25*, 1075–1082.
Meyer-Höfer, M. V., & Spiller, A. (2013). *Anforderungen an eine nachhaltige Land- und Ernährungswirtschaft: Die Rolle des Konsumenten*. KTBL-Schrift 500 Steuerungsinstrumente für eine nachhaltige Land- und Ernährungswirtschaft – Stand und Perspektiven, KTBL-Tagung vom 10-11.04.2013 in Neu-Ulm. Kuratorium für Technik und Bauwesen in der Landwirtschaft e.V. (Ed.). Darmstadt.
Reisch, L. A. (2010). *A definition of "sustainable food consumption"*. Retrieved June 15, 2015, from http://www.scp-knowledge.eu/sites/default/files/knowledge/attachments/KU_Definition_Sustainable_Food.pdf
REWE. (2014a). *Nachhaltigkeit*. Retrieved June 17, 2015, from http://www.rewe-group.com/nachhaltigkeit
REWE. (2014b). *PRO PLANET-Label*. Retrieved June 17, 2015, from https://www.rewe.de/nachhaltigkeit/gruene-produkte/pro-planet.html
REWE. (2014c). *Fragen und Antworten zum PRO PLANET-Label*. Retrieved June 17, 2015, from http://www.proplanet-label.com/fragen-und-antworten-zum-pro-planet-label
Sterzing, A. (2014). *Verteilungspräferenzen beim Kauf fair gehandelter Produkte – Eine empirische Untersuchung*. Dissertation, Kaiserslautern.
Thommen, J.-P. (1990). Glaubwürdigkeit als Grundlage des strategischen Managements. In E. Rühli & J. S. Krulis-Randa (Eds.), *Gesellschaftsbewusste Unternehmenspolitik – "Societal Strategy": Jubiläumsband zum Anlass des zwanzigjährigen Bestehens des IfbF* (pp. 121–149). Bern: Haupt.
TradeDimensions. (2013). *Die TOP 30 Food-Unternehmen des Lebensmittelhandels 2012*. Retrieved June 17, 2015, from http://www.markenartikel-magazin.de/fileadmin/user_upload/medien/Grafiken_Tabellen/LEH_Top_30_nach_Food-Umsatz.pdf
Unterbusch, B. (2011). Nachhaltigkeit in der Markenführung: Implikationen für National Brand, Private Label und Retail Brand. In E. Fröhlich, T. Weber, & C. Willers (Eds.), *Nachhaltigkeit in der unternehmerischen Supply Chain* (pp. 206–223). Cologne: Fördergesellschaft Produktmarketing.
Webb, D. J., & Mohr, L. A. (1999). A typology of consumer responses to cause-related marketing: From skeptics to socially concerned. *Journal of Public Policy and Marketing, 17*(2), 226–238.
Willers, C. (2007). *Marketing in Widerstandsmärkten – untersucht am Beispiel gentechnisch veränderter Lebensmittel*. Beiträge zum Produktmarketing (Bd. 40). Cologne.
Willers, C. (2015). Nachhaltige Produkte als Teil der Sortimentspolitik – Auswirkungen für den Lebensmittelhandel. In M. Knoppe (Ed.), *CSR & Retail Management. Gesellschaftliche Verantwortung als zukünftiger Erfolgsfaktor im Handel* (pp. 225–242). Wiesbaden: Gabler-Springer.
Willers, C., & Weber, T. (2011). Issues der Nachhaltigkeit – Unternehmen der Milchwirtschaft in der öffentlichen Wahrnehmung. *Deutsche Molkerei Zeitung*, 23/2011, pp. 32–34.
Willmann, B. (2013). Boom des guten Gewissens. In A. Kurtz (Ed.), *Factbook Einzelhandel 2012: Daten, Fakten, Trends, Perspektiven, Wirtschaftsfaktor Einzelhandel* (pp. 35–36). Neuwied: LPV Media.
Wiswede, G. (2000). *Einführung in die Wirtschaftspsychologie* (3rd ed.). Munich: Reinhardt.

The Importance of Gold in the Financial Report

Rute Abreu and Carlos Pinho

1 Introduction

The Central Banks have undergone a significant change in the last years as result of the adaptation to European Union directives to the financial report, despite the use of specific accounting system. As Zidaru (2010, p. 3) defends:

> the banking system has always been the main pillar of every economy. But more importantly, central banks have never been more powerful as the monetary policy has long become the central tool of macroeconomic stabilization.

So, the authors agree with Markwat et al. (2009) that there is a

> domino effect that causes local crises to propagate to regional or even global level, threatening the stability of the financial system. Such contagion dynamics present a strong motivation for investors to seek out a safe haven asset in times of financial crisis.

Also, for one side, Bushman (2014, p. 385) defends the role played by managerial discretion over accounting decisions in influencing bank stability through two distinct accounting channels and their interactions:

> over the accounting numbers as quantitative inputs into numerical calculations. (. . .)

R. Abreu (✉)
Guarda Polytechnic Institute and UDI-IPG and CICF-IPCA, Guarda, Portugal
e-mail: ra@ipg.pt

C. Pinho
Aveiro University, Aveiro, Portugal
e-mail: cpinho@ua.pt

H. Lu et al. (eds.), *Building New Bridges Between Business and Society*, CSR, Sustainability, Ethics & Governance, DOI 10.1007/978-3-319-63561-3_6

For the other side, Bushman (2014, p. 385) defends the bank transparency has

> a fundamental corporate governance role, because it mitigates indiscriminate panic and rollover risk by reducing depositors' and other short term lenders' uncertainty about the solvency of individual banks

In this context, the citizens recognize that the central banks influence through its actions the external environment and then it must produce effects in their economic decisions (Chua et al. 1990) and they discusses the golden dilemma (Erb and Campbell 2013). Consequently, Neuenkirch (2012, p. 1) uses

> transparency index as well as a measure of central bank speeches for a sample of nine major central banks over the period 1999 to 2007 to investigate the impact of central bank transparency and internal communication on the formation of money market expectations.

Expectations, uncertainty, problems and risk level are consequences of this complex monetary system. Also, it rises due to the lack of understandable and factually accurate accounting information and disclosure at financial report that must be made in time and constantly up-to-date to provide relevant decision to the society and not excessive controlled by the centralized state.

In this paper, the authors describe the relationship, in general, between accounting and an effective corporate governance system, achieved through recommendations and reports that it should lead main stakeholders to assume a higher degree of responsibility with lower level of risk management with relation with gold as an asset (Govett and Govett 1982). This new era obliges to change the governance of Central Bank and to understand the importance of the gold to the sustainability of the monetary system (Hillier et al. 2006). Therefore, the authors will answer to research question:

> Why gold are important to the financial report of the Central Bank?

The remainder of the article is organized as follows. In Sect. 2, it provides a literature review about the gold as an asset. Section 3 presents the methodology used to manage the documents, statistics and the data and some stylized facts. Section 4 presents the data-based evidence of the gold inventories on the European Central Bank and the Portuguese Central Bank as case studies to answer to the research question. The final section of the article summarizes the results and suggests some possible avenues for future research.

2 Literature Review

Before the literature review, it is important to define as the United Nations and European Central Bank (2014, p. 248) published on the Handbook of National Accounting, two separate definitions of gold. First the monetary gold that it is

> a component of foreign reserves is treated a s monetary gold; essentially only monetary authorities or international financial organizations hold monetary gold as a financial asset.

The definition of non-monetary gold:

> as a non-financial asset, either as a type of inventory or as a valuable; full outright ownership of the gold.

The literature review pointed out to storage theory commodities presuming that investors of inventories receive implicit benefits, called the convenience yield, but due to several economic factors (Burke 2005; Areala et al. 2013). Indeed, normally, it could decline as the level of inventory increases (Kaldor 1939; Working 1949; Brennan 1958). This result has been demonstrated by Artis et al. (1998, p. 1824) which argue that central banks and policy makers in general should say what they do and do what they say

> whole slew of reasons—not least stability, transparency and democratic accountability.

Another example, Kirkpatrick (2009, p. 1) argues that

> the financial crisis can be to an important extent attributed to failures and weaknesses in corporate governance arrangements which did not serve their purpose to safeguard against excessive risk taking in a number of financial services companies. Accounting standards and regulatory requirements have also proved insufficient in some areas.

Undeniably, in the literature about the central bank sector has the main objective to maximize benefits enough to promote social responsibility practices to the society, then it allows to develop their activity concerned with all stakeholders (EC 2011). From this perspective, Hutchison and McDill (1999, p. 160) argue that:

> a "dependent" central bank closely aligned with the government may be more inclined to provide monetary finance to problem institutions, thereby creating an additional channel for the moral hazard problem.

Obviously, the increase of laws, norms and regulations empowering the protection on the well-being of the citizen (Heliell and Putman 2004) as consequence of evolving hybrid role of accounting that it is associated with the social responsibility disclosure made on the annual report. Moreover, Resche (2015) with convincing argument reassure that:

> Central bankers, whose decisions can affect millions of people, know that they are accountable for their actions and decisions, and that their speeches will be carefully analyzed by specialists and the media, and made available to a very large public.

For all these reasons, the authors agree with Wang and Lee (2011), that defend gold has served as a global currency for a long time for its value-preserving ability. Further, the importance on the history in human affairs as the ownership of gold is likely to have many psychological dimensions (Aggarwal and Lucey 2007). Gold is also an asset (Chapman 2015; Angstrom 2009; Kaushish 2008).

3 Methodology

The methodology used on the research demonstrates the most important benefits of the financial report of the Central Banks. This research is based on the fundamental inventory of the Central Banks that is gold. So, the first part of the research presents a comparative synopsis of the organizational and legal issues related with the Central Banks at the same time it will be based on critical dimensions presented by OECD (2011a, b) promoting socially accountable finance, because as Aras and Crowther (2009) detail the society must use no more of a resource than can be regenerated.

The second part of the research is centered on the annual report of European and Portuguese Central Bank, since 1999 till 2015, which provide the accounting information system. This paper presents a exploratory and longitudinal case study examining why and how financial report integrate gold issues. In particular, it attends to these changes dynamics and their effects (Yin 2009). Likewise, Cooper and Morgan (2008, p. 165) defend that case studies can enhance research and help understand complex accounting and financial report.

The third part of the research presents findings that are consistent with the accounting and financial report literature and establish the results of Gold management according to new public management (IMF 2011; Jaffe 1989). These results show that there are an excessive financial focus, however, may adversely influence some of the more traditional activities and the promotion of best practices and the overall strategy of accounting policy. Also, the findings are consistent with the pricing mechanism of the London Gold Fix which give information to the clients that allow to buy and sell gold (Fertig 2000, p. 15).

ICE Benchmark Administration (IBA) became the administrator of the London Bullion Market Association (LBMA) that promotes an auction that takes place twice daily, by at 10:30 and 15:00, with the main objective to establish the price per fine troy ounce set in US dollars. Figure 1 shows that the London gold fix pricing mechanism has been replaced by a new electronic LBMA price-discovery process with ten price participants who have been accredited to contribute to the LBMA

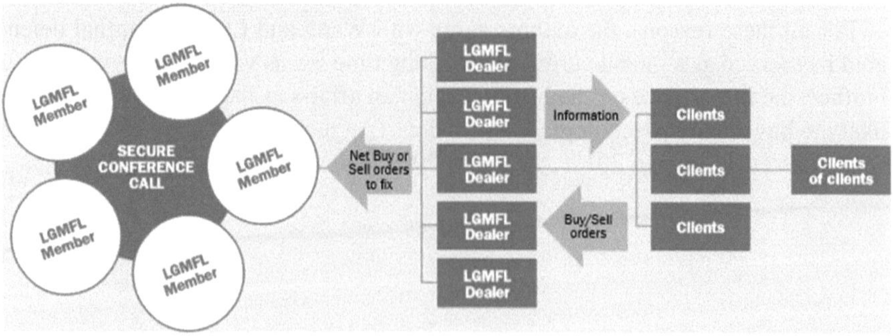

Fig. 1 The London Gold Fix pricing mechanism. Source: Fertig (2000: 15)

Gold Price, such as: *Barclays Bank, Bank of China, Goldman Sachs International, HSBC Bank USA NA, JP Morgan, Morgan Stanley, Societe Generale, Standard Chartered, The Bank of Nova Scotia—ScotiaMocatta and UBS.*

In order to generate more accountability and present increase transparency which mitigate the risk level then the United Nations and European Central Bank (2014, p. 242) published the Handbook of National Accounting where detail the definition of gold bullion:

> as a component of monetary gold held by monetary authorities as a reserve asset is treated as a financial asset even though the holder does not have claims on other designated units. There is no matching liability for gold bullion.

Despite the Bank of Portugal manages some trusts, for example, namely: the Deposit Guarantee Fund, they have been excluded from the balance sheet. So, it covers solely the activity performed by Bank of Portugal. Indeed, the balance sheet of Bank of Portugal follows the data requirements presented in the Regulation (EU) n° 1071/2013 of the European Central Bank, of 24 September 2013, concerning the balance sheet of the monetary financial institutions sector (ECB 2013b). The balance sheet includes solely the positions within Bank, which also includes the metallic coins issued by the Treasury with the contra-entry in remaining assets.

So, Governments develop and implement a monetary police that represents an efficient national instrument of solidarity of the society with the main objective of economic growth but in a sustainable way. But, these policies are directly linked to European Central Bank that promotes changes at economic level with technological improvements and redistribution of stable economic decisions.

4 Result Analysis

The Organic Law of Bank of Portugal has been published by the Law n° 5/98 of 31 January (AR 1998), as amended by Decree-Law n° 118/2001 of 17 April (MF 2001), n° 50/2004 of 10 March (MF 2004), n° 39/2007 of 20 February (MFAP 2007), n° 31-A/2012 of 10 February (MF 2012) and n° 142/2013 of 18 October (MF 2013). So, it explicitly states in the §1 of the article 13° that

> shall ensure the collection and compilation of the monetary, financial, foreign exchange and balance of payments statistics, particularly, within the scope of its co-operation with the European Central Bank.

So, the result analysis of this research demonstrates the data-based evidence since 1999 till 2015 of the Central Banks. It demonstrates that the economic crisis has encouraged the gold holdings of the central banks and better disclosure rules make information more specific and consistent.

The result analysis is justified by the opinion of Baur and McDermott (2010), when they conclude that gold thus has the potential to act as a stabilizing force for the global financial system by reducing losses when it is most needed. In particular,

Table 1 World official gold holdings, 2015–2011

Countries, June of 2015 (LPM = $1180,25)	Gold holdings Tonnes	% of the reserves	Countries, November of 2011 (LPM = $1.620)	Gold holdings Tonnes	% of the reserves
1. United States	8133.50	73.7	1. United States	8133.50	75.5
2. Germany	3383.40	67.6	2. Germany	3401.00	72.6
3. International Monetary Fund	2814.00	n.a.	3. International Monetary Fund	2814.00	n.a.
4. Italy	2451.80	66.5	4. Italy	2451.80	72.2
5. France	2435.40	65.6	5. France	2435.40	71.0
6. Russia	1246.60	13.3	8. Russia	851.50	8.6
7. China	1054.10	1.1	6. China	1054.10	1.7
8. Switzerland	1040.00	6.8	7. Switzerland	1040.10	14.3
9. Japan	765.20	2.3	9. Japan	765.20	3.3
10. Netherlands	612.50	56.9	10. Netherlands	612.50	61.3
11. India	557.70	6.1	11. India	557.70	9.3
12. Turkey	513.00	16.1	30. Turkey	116.10	6.5
13. European Central Bank	504.80	26.2	12. European Central Bank	502.10	35.0
14. Taiwan	423.60	3.8	13. Taiwan	423.60	5.7
15. Portugal	382.50	69.4	14. Portugal	382.50	88.9

Source: Adaptation of the World Gold Council (2015, 2011)

use the same metric measure instead of market prices that could be affected by markets as it is presented in Table 1.

Table 1 presents an evolution, between June of 2015 and November of 2011, of the World Official Gold Holdings with a total of 31,949.3 tonnes with an increase of 1240.7 tonnes or 4.04% of the total of 2011. In the Euro Area, including the European Central Bank, with a total of 10,792.5 tonnes and several countries, like Italy, France, Netherlands and Portugal keep the same volume of gold as an asset of their international reserve portfolios and this implies that changes are monitored by markets, investors, policymakers and producers.

The biggest increase was Turkey with more 396.9 tonnes or 341.9% of the total of 2011. The Germany has decrease with small reduction of 17.6 tonnes or 0.5% of the total of 2011. This data, also, confirmed the results of the research developed by Shafiee and Topal (2010) that present the gold demand to jewelry, industrial and central bank reserves equate to approximately 100,000, 30,000 and 30,000 tonnes, respectively, and the gold has been mined in history up to the end of 2008 is around 160,000 tonnes. Table 2 presents the information detailed in the annual report related with gold and gold receivables that are located at the European Central Bank.

The obvious conclusion of Table 2 is the inverse relation between the increase value of the Gold and Gold receivables on the balance sheet of the ECB and the decrease of the unit per troy ounce. Yet this shows the increase of the price that ironically distorted the power of this information. Also, Table 2 proves that the

Table 2 Gold asset of the balance sheet of European Central Bank, 1999–2014

Year (31 December)	Balance sheet of ECB			London Gold Monetary Fixing (LGMF)	
	Gold and gold receivables (€)	Quantity per troy ounce (€)	Quantity per tonne (€)	Price per troy ounce (US $)	Price per troy ounce (€)
1999	6,956,995,273	24,000,000	680.39	290.25	289.56
2000	7,040,906,565	24,000,000	680.39	274.45	292.31
2001	7,766,256,040	24,700,000	700.23	276.50	310.53
2002	8,058,187,254	24,700,000	700.23	347.20	330.86
2003	8,145,320,117	24,700,000	700.23	416.25	330.00
2004	7,928,308,842	24,700,000	700.23	435.60	320.47
2005	10,064,527,857	23,100,000	654.87	513.00	434.91
2006	9,929,865,976	20,572,017	583.21	632.00	479.28
2007	10,280,374,109	18,091,733	512.89	833.75	570.26
2008	10,663,514,154	17,156,546	486.38	869.75	625.70
2009	12,355,158,122	16,122,146	457.06	1087.50	757.97
2010	17,015,600,109	16,122,143	457.06	1405.50	1047.67
2011	19,643,678,205	16,142,871	457.64	1531.00	1179.37
2012	20,359,049,520	16,142,871	457.64	1657.50	1257.21
2013	14,063,991,807	16,142,871	457.64	1204.50	874.12
2014	15,980,317,601	16,178,193	458.65	1206.00	996.65

Source: Adaptation of the ECB (2000, 2001, 2002, 2003, 2004, 2005, 2006, 2007, 2008, 2009, 2010, 2011, 2012, 2013a, 2014, 2015), LMBA (2015)

highest price (in Euro) has been obtained in year 2010 when compare with year 1999 around 265% or 767,36 € per troy ounce and lowest price (in Euro) has been obtained in year 2000 compared with the year 1999 around 2% or 6,49 € per troy ounce. In particular, the last 3 years (2008–2010) have seen the price of gold rise up to 100%. And the last information available the price is increasing till 1.222,86 € or 323% versus $1594,00 or 449% when compare with 1999.

The gold recent history may be show disproportionate speculation due to unusual economic uncertainty, and then the authors agree with Graafland and van de Vem (2011) when they defend the urgent need of the regulation of the financial sector to prevent negative externalities of banking on other sectors of economy. On 2014 happens an increase due to the transfer made by the Latvijas Banka to the ECB of 35,322 ounces with value equivalent to 30.476.837 €, of fine gold upon the adoption of the single currency by Latvia. Indeed the increase of gold and gold receivables were bigger due to the changes of 122,53 € per troy ounce from 2013 to 2014.

Table 3 presents the information detailed in note 2 of the Annual Report of Bank of Portugal related with gold and gold receivables that are located in Portugal and abroad, which location are: the Bank of England, Bank for International Settlements (BIS) and Federal Reserve Bank on United States (FED).

Table 3 Gold asset of the balance sheet of Bank of Portugal, 1999–2014

Year (31 December)	Balance sheet of Bank of Portugal					
	Gold and gold receivables (€) Portugal	Gold and gold receivables (€) Abroad	Quantity per troy ounce-Portugal	Quantity per troy ounce-Abroad	Price per troy ounce (US $)	Price per troy ounce (€)
1999	1,607,155	4,039,953	172,659,668	434,019,839	290.85	289.52
2000	1,626,514	4,088,940	172,657,096	434,047,538	272.65	293.01
2001	1,748,527	4,395,849	172,657,096	434,064,988	277.60	314.99
2002	1,814,250	4,404,651	172,657,096	419,178,368	326.83	326.83
2003	1,833,869	3,659,118	172,657,096	344,502,741	417.25	330.36
2004	1,785,008	2,994,209	172,657,096	289,618,531	438.00	321.56
2005	2,413,909	3,422,635	172,657,097	244,807,126	513.00	434.86
2006	2,679,427	3,257,457	172,657,097	209,904,181	636.00	482.69
2007	3,154,309	3,834,652	172,657,097	209,897,028	837.00	568.24
2008	3,450,147	4,194,114	172,625,844	209,883,708	865.00	621.54
2009	4,253,264	5,171,246	172,653,775	209,883,739	1104.00	766.35
2010	5,857,622	7,121,872	172,625,844	231,081,601	1410.00	1055.42
2011	6,752,863	8,211,296	172,605,595	292,831,840	1574.50	1216.86
2012	6,998,784	5,298,530	172,605,595	209,883,677	1664.00	1261.18
2013	4,834,746	5,878,920	172,605,595	209,883,677	1201.50	871.22
2014	5,481,372	6,665,382	172,600,836	209,883,677	1199.25	987.77

Source: Adaptation of the BP (2000, 2001, 2002, 2003, 2004, 2005, 2006, 2007, 2008, 2009, 2010, 2011, 2012, 2013, 2014, 2015)

Tables 2 and 3 present a difference between the European Central Bank and Portuguese Central Bank that is the rate at which the bank made exchange of the gold to Euro and US Dollar. The difference could be justified for inclusion of the cost of transactions. It is important to notice that the changes of the value of the gold are related with the accounting process (BP 2015):

> revaluation differences associated with the gold reserve are calculated taking as a basis the price of the ounce of fine gold denominated in euro, and no distinction is made between the gold price revaluation and the foreign currency revaluation.

On 2014, the Portuguese Central Bank's gold reserve was valued at the market price of 987.77 € per ounce of fine gold, which corresponds to a 241.2% price rise, from 289.52 € on 31 December 1999. The increase in this price was due to the appreciation of the US dollar against the euro (14.0%) between these two dates. During the period of 1999 till 2006 the average of the price is 191.5% and the standard deviation of 49.5%. As Aggarwal and Soenen (1988) defend the transactions in the gold market by central banks are generally not characterized by profit maximizing behavior and it is possible to observe there are not transactions. Indeed, this situation face a very important turn around during the years 2010 till 2012 with negatives values of reduction through 15.6% of average and 8.1% of standard deviation.

Logically, Laeven (2013) has also evidenced that central banks must promote discussions of what makes banks special in terms of corporate governance. Likewise, Ford (1989, p. 202) argue that

> against accepting too easily some of the facile stereotypes which have been applied to gold standard provisions regardless of individual circumstances' thus continues to stand.

5 Conclusion

This research proves the importance of gold as public asset (Kindleberger 1981), but due to the demand made by citizens and invertors then it must be paid to be private asset (Richins 1994; Sari et al. 2010). Indeed, there are rules of the gold standard used by Central Banks that they were far from universal. This is only possible to understand when the authors are aware of the study of these institutions in profundity (Vogel 2005). In addition, the results show the need of more transparency and the increase awareness of the communication process of the real state of the economy and the annual report of the Central Banks (European and Portuguese) as an attitude of willingness to learn from past mistakes and then they must cooperate with Society to promote the economic growth.

For the Economic System, where the producers as economic agents influence markets due to the production and selling process that allows gold to be available at markets with best price. The citizen lives in the environment facing social and economic rules of gold circulation.

For the Financial System, the investor's must develop investments in areas of free of conflict of interest and promote gold supply chains at national and international level and guarantee the accountability of these companies with respect to the implementation of accounting and auditing international standards.

For the Political System, there is a growing commitment of Governments to planning and approving laws and regulation that control the several marketing tendencies, such as: speculation, illicit trade without paying taxes, as well as, fighting power suppliers that it will emerging responsible, ethical, social and environmental good practices. For national and international agencies that publish guidance that comply with high standards of anti-money laundering and combating systems that finance miss practice all over the world.

For the Education System, professors must teach the financial stability that protects against threats to the global financial system using surveillance and ICT functions. Indeed, the accounting and monetary systems promotes the stable price and confidence on the currency in order to improve the economic stability. Students and researchers must study the evolution of markets and accounting systems in order to achieve a sustainable future evolution . . . and this will be seen in the future!

For possible avenues on the future research include the significant analysis of the increase in the level of interest of researchers, regulators, legislators, citizens and broader groups surrounding corporate governance of entities and national agencies that control the gold market. Obviously, the argument for analyzing it through the lens of gold management recently started receiving serious attention. More often, the future suggestion will help to clarify the externalities of the specific context of each Central Bank and in the quantitative research design examining constructs (or variables) that were included in your conceptual framework (or theoretical model) but were not focused on the literature.

Acknowledgment The authors wish to thank José Ángel Pérez López of Universidad de Seville (Spain). Ideas expressed in the article are those of the authors and should not be attributed to any organization.

References

Aggarwal, R., & Lucey, B. (2007). Psychological barriers in gold prices? *Review of Financial Economics, 16*(2), 217–230.
Aggarwal, R., & Soenen, L. (1988). The nature and efficiency of the gold market. *Journal of Portfolio Management, 14*(3), 18–21.
Angstrom, H. (2009, February 26). *Haring away*. London: The Economist Newspaper Limited.
Aras, G., & Crowther, D. (2009). Corporate sustainability reporting: A study in disingenuity? *Journal of Business Ethics, 87*(3), 279–288.
Areala, N., Oliveira, B., & Sampaio, R. (2013). When times get tough, gold is golden. *The European Journal of Finance, 21*(6), 507–526.
Artis, M., Mizen, P., & Kontomelis, Z. (1998, November). Inflation targeting: What can the ECB learn from the recent experience of the Bank of England? *The Economic Journal, 108*, 1810–1824.
Assembleia da República (AR). (1998, January 31). Law n° 5/98. Diário da República, 1ª Série A, n° 26.
Banco de Portugal (BP). (2000). *Relatório do Conselho de Administração. Atividades e Contas 1999*. Lisboa: BP.
Banco de Portugal (BP). (2001). *Relatório do Conselho de Administração. Atividades e Contas 2000*. Lisboa: BP.
Banco de Portugal (BP). (2002). *Relatório do Conselho de Administração. Atividades e Contas 2001*. Lisboa: BP.
Banco de Portugal (BP). (2003). *Relatório do Conselho de Administração. Atividades e Contas 2002*. Lisboa: BP.
Banco de Portugal (BP). (2004). *Relatório do Conselho de Administração. Atividades e Contas 2003*. Lisboa: BP.
Banco de Portugal (BP). (2005). *Relatório do Conselho de Administração. Atividades e Contas 2004*. Lisboa: BP.
Banco de Portugal (BP) (2006). *Relatório do Conselho de Administração. Atividades e Contas 2005*. Lisboa: BP.
Banco de Portugal (BP). (2007). *Relatório do Conselho de Administração. Atividades e Contas 2006*. Lisboa: BP.
Banco de Portugal (BP). (2008). *Relatório do Conselho de Administração. Atividades e Contas 2007*. Lisboa: BP.
Banco de Portugal (BP). (2009). *Relatório do Conselho de Administração. Atividades e Contas 2008*. Lisboa: BP.

Banco de Portugal (BP). (2010). *Relatório do Conselho de Administração. Atividades e Contas 2009*. Lisboa: BP.
Banco de Portugal (BP). (2011). *Relatório do Conselho de Administração. Atividades e Contas 2010*. Lisboa: BP.
Banco de Portugal (BP). (2012). *Relatório do Conselho de Administração. Atividades e Contas 2011*. Lisboa: BP.
Banco de Portugal (BP). (2013). *Relatório do Conselho de Administração. Atividades e Contas 2012*. Lisboa: BP.
Banco de Portugal (BP). (2014). *Relatório do Conselho de Administração. Atividades e Contas 2013*. Lisboa: BP.
Banco de Portugal (BP). (2015). *Relatório do Conselho de Administração. Atividades e Contas 2014*. Lisboa: BP.
Baur, D., & McDermott, T. (2010). Is gold a safe haven? International evidence. *Journal of Banking & Finance, 34*(8), 1886–1898.
Brennan, M. (1958). The supply of storage. *American Economic Review, 48*, 50–72.
Burke, E. M. (2005). *Managing a company in an activist world: The leadership challenge of corporate citizenship*. Westport, CT: Praeger.
Bushman, R. (2014). Thoughts on financial accounting and the banking industry. *Journal of Accounting and Economics, 58*(2/3), 384–395.
Chapman, J. (2015). *Emotionally durable design: Objects, experiences and empathy*. New York: Routledge.
Chua, J. H., Sick, G., & Woodward, R. S. (1990). Diversifying with gold stocks. *Financial Analysts Journal, 46*(4), 76–79.
Cooper, D., & Morgan, W. (2008). Case study research in accounting. *Accounting Horizons, 22*(2), 159–178.
Erb, C. B., & Campbell, R. H. (2013). The golden dilemma. *Financial Analysts Journal, 69*(4), 10–42.
European Central Bank (ECB). (2000). *Annual report of 1999*. Frankfurt: ECB.
European Central Bank (ECB). (2001). *Annual report of 2000*. Frankfurt: ECB.
European Central Bank (ECB). (2002). *Annual report of 2001*. Frankfurt: ECB.
European Central Bank (ECB). (2003). *Annual report of 2002*. Frankfurt: ECB.
European Central Bank (ECB). (2004). *Annual report of 2003*. Frankfurt: ECB.
European Central Bank (ECB). (2005). *Annual report of 2004*. Frankfurt: ECB.
European Central Bank (ECB). (2006). *Annual report of 2005*. Frankfurt: ECB.
European Central Bank (ECB). (2007). *Annual report of 2006*. Frankfurt: ECB.
European Central Bank (ECB). (2008). *Annual report of 2007*. Frankfurt: ECB.
European Central Bank (ECB). (2009). *Annual report of 2008*. Frankfurt: ECB.
European Central Bank (ECB). (2010). *Annual report of 2009*. Frankfurt: ECB.
European Central Bank (ECB). (2011). *Annual report of 2010*. Frankfurt: ECB.
European Central Bank (ECB). (2012). *Annual report of 2011*. Frankfurt: ECB.
European Central Bank (ECB). (2013a). *Annual report of 2012*. Frankfurt: ECB.
European Central Bank (ECB). (2013b, November 7). Regulation (EU) n° 1071/2013 of 24 September. *Official Journal*, L297.
European Central Bank (ECB). (2014). *Annual report of 2013*. Frankfurt: ECB.
European Central Bank (ECB). (2015). *Annual report of 2014*. Frankfurt: ECB.
European Commission (EC). (2011, October 25). *A renewed EU strategy 2011–14 for corporate social responsibility, COM (2011) 681 final*. Brussels: Official publications of the European Commission.
Fertig, P. (2000). Has there been a decade of London PM gold fixing manipulation? *Alchemist Issue, 73*, 14–15.
Ford, A. G. (1989). International financial policy and the gold standard, 1870–1914. In P. Mathias & S. Pollard (Eds.), *The Cambridge economic history of Europe, The industrial economies:*

The development of social and economic policies (Vol. viii). Cambridge: Cambridge University Press.
Govett, M. H., & Govett, G. J. S. (1982). Gold demand and supply. *Resources Policy, 8*(2), 84–96.
Graafland, J., & van de Vem, B. W. (2011). The credit crisis and the moral responsibility of professionals in finance. *Journal of Business Ethics, 103*(4), 605–619.
Heliell, J., & Putman, R. (2004). The social context of well-being. *Philosophical Transactions of the Royal Society of London, Serie B, 359*, 1435–1446.
Hillier, D., Draper, P., & Faff, R. (2006). Do precious metals shine? An investment perspective. *Financial Analysts Journal, 62*(2), 98–106.
Hutchison, M. M., & McDill, K. (1999). *Are all banking crises alike? The Japanese experience in international comparison*. Federal Reserve Bank of San Francisco, Center for Pacific Basin Studies Working Paper PB99-02.
International Monetary Fund (IMF). (2011). *Gold in the IMF. Factsheet – September*. New York: IMF.
Jaffe, J. F. (1989). Gold and gold stocks as investments for institutional portfolios. *Financial Analysts Journal, 45*(2), 53–59.
Kaldor, N. (1939). Speculation and economic stability. *Review of Economic Studies, 7*, 1–27.
Kaushish, J. P. (2008). *Manufacturing processes*. New Delhi: Eastern Economy Edition.
Kindleberger, C. P. (1981). Dominance and leadership in the international economy: Exploitation, public goods, and free rides. *International Studies Quarterly, Symposium in Honor of Hans J. Morgenthau, 25*(2), 242–254.
Kirkpatrick, G. (2009). The corporate governance lessons from the financial crisis. *Financial Markets Trends, 2009*(1), 1–30.
Laeven, L. (2013). Corporate governance: What's special about banks? *Annual Review of Financial Economics, 5*, 63–92.
London Bullion Market Association (LBMA). (2015). *Gold fixings – historical statistics*. London: LBMA.
Markwat, T., Kole, E., & van Dijk, D. (2009). Contagion as a domino effect in global stock markets. *Journal of Banking and Finance, 33*, 1996–2012.
Ministério das Finanças (MF). (2001, April 17). Decree-Law n° 118/2001. *Diário da República*, 1ª Série A, n° 90.
Ministério das Finanças (MF). (2004, March 10). Decree-Law n° 50/2004. *Diário da República*, 1ª Série A, n° 59.
Ministério das Finanças (MF). (2012, February, 10). Decree-Law n° 31-A/2012. *Diário da República*, 1ª Série A, n° 30, Supplement.
Ministério das Finanças (MF). (2013, October 18). Decree-Law n° 142/2013. *Diário da República*, 1ª Série A, n° 202.
Ministério das Finanças e da Administração Pública (MFAP). (2007, February 20). Decree-Law n° 39/2007. *Diário da República*, 1ª Série A, n° 36.
Neuenkirch, M. (2012). Managing financial market expectations: The role of transparency and central bank communication. *European Journal of Political Economy, 28*(1), 1–13.
OECD. (2011a). *OECD due diligence guidance on responsible supply chains of minerals from conflict-affected and high-risk areas*. Paris: OECD.
OECD. (2011b). *Supplement on gold – final draft*. Paris: OECD.
Resche, C. (2015). Hedging in the discourse of Central Banks. *Studies in Communication Sciences, 15*(1), 83–92.
Richins, M. L. (1994). Valuing things: The public and private meanings of possessions. *Journal of Consumer Research, 21*(3), 504–521.
Sari, R., Hammoudeh, S., & Soytas, U. (2010). Dynamics of oil price, precious metal prices, and exchange rate. *Energy Economics, 32*, 351–362.
Shafiee, S., & Topal, E. (2010). An overview of global gold market and gold price forecasting. *Resources Policy, 35*(3), 178–189.

United Nations and European Central Bank. (2014). *Handbook of national accounting*. New York: UN.
Vogel, D. (2005). *The market for virtue*. Washington, DC: Brookings Institution Press.
Wang, K.-M., & Lee, Y.-M. (2011). The yen for gold. *Resources Policy, 36*(1), 39–48.
Working, H. (1949). The theory of the price of storage. *American Economic Review, 39*, 1254–1262.
World Gold Council. (2011, November). World official gold holdings. *International Finance Statistics*. London: WGC.
World Gold Council. (2015, June). World official gold holdings. *International Finance Statistics*. London: WGC.
Yin, R. K. (2009). *Case study research: Design and methods*. Thousand Oaks, CA: Sage.
Zidaru, S. (2010). *The European Central Bank: Monetary policy instruments, crisis management and European macroeconomic performane*. Msc Thesis, Finance and International Business.

Part III
CSR and Management

Sustainable Hospitality Management: Challenges and Opportunities for Small Island Destinations—Lessons from the British Virgin Islands

Bonnie Lewtas

List of Abbreviations

BVI	British Virgin Islands
IRF	Island Resources Foundation
SDM	Sustainable Destination Management
SHM	Sustainable Hospitality Management
SIDs	Small Island Destinations
SIDS	Small Island Developing States
SIDHs	Small Island Destination Hotels
WM	Waste Management

1 Introduction

> I think we have done so well here because of the way we do things when it comes to the environment (Hotel B 2015)

This research determines sustainability challenges for small island destination hotels (SIDHs) and how obstacles can be overcome or avoided through innovative and integrative management techniques. It should aid hotels in the establishment of an efficient long-term stable business approach. This paper is not meant to offer all the solutions to every challenge faced by hotels. Rather, it develops a broad understanding of sustainable hospitality management (SHM), why it is important and areas of consideration. Various challenges and solutions are discussed referencing a selection of examples derived from academic and onsite research. Key areas identified as most important for SIDHs are energy, water, waste wildlife conservation and food and beverage. Not included are the categories: Building and construction,

B. Lewtas (✉)
Cologne Business School (CBS), Cologne, Germany
e-mail: bonnie.lewtas@gmail.com

H. Lu et al. (eds.), *Building New Bridges Between Business and Society*, CSR, Sustainability, Ethics & Governance, DOI 10.1007/978-3-319-63561-3_7

purchasing of building materials, sustainable transportation, guest activities, and the social dimension of sustainability. Due to the holistic nature of the topic overlap between categories is to be expected. The overlay between SHM and sustainable destination management (SDM) makes this paper informative for destination planners and decision makers alike. The sustainable management of SIDHs is diverse and, despite the limited scope of this research, the amount of management areas, challenges and possible solutions provided may seem overwhelming for hoteliers. However, sustainability management is at the core of long-term business success and every initiative undertaken directly or indirectly supports the long-term viability of hotels.

2 Methodology

This paper was derived from information submitted to Cologne Business School (CBS), Cologne, Germany as a master's thesis in the program Sustainable Tourism Management. The allotted research period was approximately four months. Theoretical foundations for this paper were developed from personal interviews with field experts, as well as sustainable hotel certification frameworks such as Green Globe, Green Key, and the Sustainable Hospitality Industry Guide by the International Tourism Partnership. Along with a literature review of global examples in the field, primary research was conducted in the British Virgin Islands using grounded theory for social research. All communications were face-to-face semi-structured interviews. Interview durations lasted between fifteen minutes and four hours. While several hotel experts did not respond to email/telephone requests, overall, the researcher experienced a notable amount of support and the majority of people contacted were keen participants. Interviewees did not seem hesitant to transparently discuss the topic and provided a lot of valuable information. Interviews were relaxed in nature and took place on-site or at local venues. Participants commonly recommended and provided connections for further interviews. Many more field experts would have been available for communications; however, due to time constraints this was not feasible. In order to maintain anonymity, hotel interviews have been labeled A, B, C and D and interviews with field experts have been given titles according to their field of expertise. A list of personal communications conducted is provided in the Appendix at the end of this paper. The exhibition at the BVI 2015 OCTA Ministerial Conference Showcasing Sustainable Development through—Competitiveness and Green Growth was also attended. There, two interviews were conducted and government representatives provided industry reports.

3 Results

According to a review of pertinent literature, there is no published paper offering a comprehensive approach to SHM specific to SIDs. Therefore, further research of this topic is of major significance for hotels to identify solutions in key categories. Factors making SHM difficult on SIDs are isolation from economies of scale, natural resource scarcity, import dependency, increasing demand, and fragile eco-systems. Key areas of focus for SHM on small islands have been identified as energy, waste, water, wildlife conservation, and food and Beverage. There exist many examples of NGO, governmental, business and local initiatives in this field and in order to maintain business, hotels must play a large role in sustainable solutions. The findings of this research have confirmed that sustainability is positively correlated with successful hotel management on SIDs.

The following is a discussion of the five key management areas identified as most critical for hotels on small islands. Examples in each category are provided through a combination of international literature and personal communications with industry experts in the British Virgin Islands (BVI) who offer an up-close perspective on the topic. This paper begins with an overview of SDM for small islands and narrows to SHM for SIDs. Following is a discussion of different challenges and possible solutions in the above-mentioned areas (energy, water, waste, wildlife conservation and food and beverage). As all primary research was conducted in the BVI, local examples are integrated into core chapters. This paper is meant to serve as a bridge between academia and business communities.

3.1 Sustainable Destination Management for Small Islands

The value of this research depends on the identification of factors unique to SIDs. Discovering why SIDHs require an adapted set of sustainability management principals from hotels in continental regions is critical for the development of the SHM discussion. Globally, SIDs have incredibly diverse cultural and geographical features. Still, they share sustainable development challenges stemming from economic and environmental vulnerabilities (Rogers et al. 2012, p. 284). In 1972, the founder and president of the BVI Island Resources Foundation (IRF) discussed the uniqueness of island regions. Small islands require a more integrated and holistic management approach as all actions have a domino effect (IRF 2012, p. 65). This is not only true of ecological effects but also between people and businesses. A hotel manager on a small island in the Caribbean discusses the need to take this into account when making management decisions: "In remote locations everything has a contributing factor and every decision that you make has a knock on effect on probably ten other people (p.c., Hotel B 2015, p. 28)." Explained was the importance of considering the impact of each option and trying to choose the path with the most positive effects on its interdependent relationships (p.c., Hotel B 2015, p. 28).

Despite the negative environmental impacts of tourism development, it is the main economic pillar for many small islands (Varela-Acevedo et al. 2009, p. 3). If replaced by more environmentally friendly activities much economic hardship would occur (UNWTO 1998, p. 698). If properly managed, SHM should positively contribute to the environmental, social and economic prosperity of the region (UNWTO 1998, p. 700). This is not an easy task. The challenges facing islands are unprecedented (Purkis and Miller 2012, p. 22). Climate change is already being felt (UNESCO 2010, p. 3). According to the Intergovernmental Panel on Climate Change (IPCC) and the United Nations Framework Convention on Climate Change (UNFCCC), small islands will be hit first and hardest from the negative affects of climate change (IRF 2012, p. 61). Consequences include rising temperatures (Varela-Acevedo et al. 2009, p. 3) (leading to changes in tourist demand), decreased overall rainfall (limiting agricultural capacity) coupled with heavier rainfall events (causing flooding and ocean run-off), more and stronger hurricanes (threatening infrastructure, wildlife and the safety of locals and tourists), as well as sea level rise (IRF 2012, p. 61) (increasing beach erosion) (CANARI 2012, p. 41). Other obstacles include overshot carrying capacity, increased air pollution, degraded water and soil quality, waste management (WM) issues (Purkis and Miller 2012, p. 22), dependence on imports, and ocean acidification from increased greenhouse gas emissions (Summit Secretariat 2013, p. 17). These factors present severe risks for the present and future quality of life on SIDs (Purkis and Miller 2012, p. 22).

If SIDs ensue strong action to protect natural resources, they will be better positioned to cope with future climate threats and the protection of their small economies (Summit Secretariat 2013, p. 17). Development sites and systems must be designed with enough flexibility to respond to potential changes and stress sources (Gardner et al. 2008, p. 83). In many cases this is not occurring. Tourism development creates a playground for visitors with easy access to the most pristine and valuable sites. Proper planning must precede development to ensure the undertaking of best management practices during construction and operation to avoid serious environmental damage (IRF 2012, p. 139). Development activities commonly move ahead without sufficient ecosystem protection measures (UNWTO 1998, p. 700). Tourism dependent small island economies tend to have heavily developed coastlines. The “Sun and Sand” trend exacerbates biodiversity and leaves the area more vulnerable to erosion (IRF 2012, p. 139). There is a positive correlation between hotel adjacency and the lowest and most narrowing beaches (Varela-Acevedo et al. 2009, p. 3). On small islands, when roads, hotels, houses, and other infrastructure develop behind a beach, there is nowhere to retreat inland and land increasingly narrows eventually disappearing. Common solutions such as the building of offshore breakwater and groynes or replenishing beach sand from offshore or land-based sources are only short-term fixes (CANARI 2012, p. 41). SIDHs should become involved with SDM to create long-term solutions to the above-mentioned challenges.

3.2 Sustainable Hospitality Management for Small Island Destinations

With tourism being the primary economic pillar of many small islands (UNWTO 1998, p. 67), and hotels being the main form of development (Prayag et al. 2010, p. 698), SHM is critical (UNWTO 1998, p. 697). The SHM discussion can be traced back to the 1960s; however there has been a significant increase in awareness over the past few decades. Now more than ever hoteliers and investors take into consideration the ecological and social impacts of hotel construction and operation (Jones et al. 2014, p. 6). According to a 2010 study of perspectives on impacts of hotel development in Mauritius, the majority of hoteliers thought that the development of hotels should positively contribute to environmental, social, and economic sustainability in the region where they operate (UNWTO 1998, p. 697). To host a successful tourism industry on a small island it is important for decision makers to consider how hotels impact the surrounding environment and community (Prayag et al. 2010, p. 698).

Sustainable hotel management requires a set of principles and management methods that allow tourism to develop in conjunction with the protection of the destination's environmental, social, and infrastructural resources (UNWTO 1998, p. 700). Funds must be directed towards sustainability projects. Currently, not nearly enough resources to maintain a viable tourism industry on SIDs are going towards sustainability (p.c., Hotel A 2015, p. 13). This can serve as a barrier for managers who see the long-term value (also economic) of putting sustainability ahead of cost cuts. "There have been places we have worked where it is totally frustrating because they won't change and they know its wrong, but they continue to do it wrong because they haven't got the capital investment to fix it, or they just don't care enough (p.c., Hotel B 2015, p. 8)." Education both internally and of external stakeholders can help direct resources to sustainability projects. Long-term solutions to the environmental and social issues faced by hotels on SIDs will come from investment in innovative sustainable development and conservation initiatives (Jones et al. 2014, p. 8).

Still, after awareness on the importance of sustainability has been created and managers are willing to act, many SHM challenges still exist. Time is a key challenge (Hotel B 2015, p. 8). Positions in the hospitality sector are very demanding and the implementation of sustainability measures can add additional work to the already stressed schedules of top employees (p.c., Hotel C 2015, p. 3). SIDHs can however no longer afford to ignore their environmental management responsibilities, as problems will become more complicated and expensive to deal with as time progresses. Climate and weather patterns are significantly changing. This compromises the comfort and safety of tourists, which will intern have an effect on their choice of destination (ITP 2014, p. 4). Whether it be decreases in occupancy, a widening gap between seasonal occupancy levels, or increased demand, hotel businesses (as well as industry partners such as suppliers) will be affected (ITP 2014, p. 4) and have to adapt management strategies accordingly. This should be particularly prepared for on SIDs where tourism, which relies on predictable climates, is an economic pillar (ITP 2014, p. 4). This demonstrates the financial interest of the hotel industry to aid in environment stabilizing solutions.

One of the key challenges facing hotel managers who want to invest in sustainability is the existence of old infrastructure, especially for hotels with a high-level of occupancy. "Every engineer's dream would be to shut down and build everything from scratch. However, because of the high level of occupancy, that will never happen here (p.c., Hotel A 2015, p .7)." Furthermore, hotel properties are often so expensive to run (regardless of occupancy levels) that it is not economically feasible to shut down and replace inefficient infrastructure (p.c., Hotel A 2015, p. 7). Therefore, hoteliers must rely on quick fixes for energy, water and waste. Managing guest expectations while trying to reduce the consumption of scarce resources represents another challenge. Most guests do not want to significantly change their lifestyles while on holiday and desire all the amenities they have at home plus more. This can contradict sustainability initiatives (p.c., Hotel C 2015, p. 9). It is especially difficult to manage at luxury resorts that cater to demanding tourists. The high cost of luxury accommodation makes guests more reluctant to limit consumption. When asked about this issue one expert answered: "It is a challenge—but it is what it is. The focus is about doing the best job achievable to make systems as efficient as possible (p.c., Hotel A 2015, p. 11)." Another hotel manager pointed out that investment in sustainability initiatives by high-end hotels can help to educate clientele in powerful positions, which can intern initiate change (p.c., Hotel B 2015, p. 31). Furthermore, high-end hotels bring in tourists who spend a lot of money on the island and, if sustainability is a priority, are more likely to get the government on board with initiatives to protect the associated economic interest (p.c., Hotel B 2015, p. 31).

While environmental initiatives will not always align with unrealistic guest expectations, sustainable hotel management does not have to mean offering a downgraded guest experience. One hotelier who has experienced many positive benefits from implementing sustainability measures discussed that the term "eco-resort" is often wrongly considered to be lacking something. To this hotelier, environmental management is a normal and integrated part of the job: "It makes not just financial sense, it makes common sense to save the environment. We are intelligent human beings, we should be able to have whatever we want, but do so in a way that has as low of an impact as possible (p.c., Hotel B 2015, p. 16)." This paper is meant to offer solutions to hoteliers' sustainability challenges that do exactly that; manage environmental resources in the most efficient and effective ways without reducing guest experience—often enhancing it. Due to the additional struggles associated with operating an accommodation business on a small island (remoteness, limited resources, fragile ecosystems, reliance on imports, etc.), hotels on SIDs should do absolutely everything in their power to effectively and efficiently manage resources. The successful hotels of the future will have managers that are constantly thinking about what can be improved to offer guests a better experience with fewer resources (p.c., Hotel B 2015, p. 21). When energy, water, land, and human resources are severely limited, or costly, the less these resources are wasted through efficient management the more will be available for a quality guest experience over a longer period of time. There is no "one size fits all" 100% effective solution to SHM. It is a process consisting of day-to-day decisions (both small and large) and reflection upon the impact of those decisions. There will always be room for improvement. As external factors change so must management strategies. In order to determine what steps to take, it is important to know where the company is with regards to resource consumption and environmental impact. Therefore, measuring impact is the first step hoteliers must take to reduce their environmental footprint (Jones et al. 2014, p. 9). The goal of hotel managers should be to achieve the highest-level of self-sufficiency (for the island—local partnerships are encouraged) with the lowest environmental footprint. Hoteliers who keep this core value in mind while making all decisions are on the path to sustainability. Solutions to sustainability challenges are cross-sectional and the division of standard hospitality management rolls may not allow for adequate action. Nevertheless, the hotel industry is experiencing a sustainability awakening, which owners and managers can no longer afford to ignore (Jones et al. 2014, p. 6). Solutions to challenges and the cross-sectional flexibility required to implement them should be internally developed in the organization.

3.3 Energy Management

Islands are highly dependent on energy imports and due to common characteristics such as small populations and isolation from large economies, are more susceptible than mainland countries to external factors such as global energy cost spikes and scarcity. Furthermore, increased transportation costs (Rogers et al. 2012, p. 284) and associated risks such as oil spillage (Boniface and Cooper 2009, p. 49) must be considered. Self-sufficiency should be the primary goal for SIDHs. The consumption of fossil fuels not only creates a dependence on oil imports but also does not make economic sense. In 2008, SIDS collectively spent over 90 million US dollars per day on 900,000 barrels of oil. This money could have yielded far higher long-term profits (Rogers et al. 2012, p. 285) and greatly contributed to self-sufficiency (Boniface and Cooper 2009, p. 49) if it were to have been invested in renewable energy technologies (Rogers et al. 2012, p. 285). Another economic benefit of switching to renewable energies is the positive impact on guest experience. Having a diesel generator running all day is loud and disturbs an otherwise relaxing atmosphere at SIDHs (p.c., Hotel A 2015).

Governments on SIDs either help or hinder the development of renewable energy systems. According to a study conducted in Barbados, most of the successful renewable energy projects have occurred on non-sovereign islands due to mainland financial incentives. The majority of SIDHs however are dependent on international bilateral and multilateral investment agreements (Rogers et al. 2012, p. 289). Still, financial incentives are on the rise and, on some islands, exist to urge businesses to switch to renewables (Rogers et al. 2012, p. 286). With the development of sustainability and Corporate Social Responsibility (CSR) initiatives in global public discourse, this is likely to continue. Unfortunately, not all hotels can take advantage of such programs. Many small islands lack the political foundation commonly demonstrated to drive sustainable technologies through supportive policy implementation (Rogers et al. 2012, p. 286). Policy related obstacles resulting from a shortage of education, research and awareness commonly affect hotels (Rogers et al. 2012, p. 289). These can be either internally within the organization or externally at the political level. Hotels on SIDs that are able to independently invest in renewable energies should demonstrate the benefits of such projects by inviting governments to the property and explaining systems and associated profitability. This helps local governments to change perspectives on renewable energies and consider adapted legislation (p.c., Hotel B 2015 p. 11).

When it comes to the development of renewable energy on small islands it is important that hotels are transparent about their initiatives and try to invest in scalable solutions as pilot projects (p.c., Hotel A 2015, p. 5). Only projects that make financial sense will encourage other businesses to invest in the same or similar technologies (p.c., Hotel A 2015, p. 6). While scalable solutions are important, it is also necessary that hotels determine the specifics of their needs and what is viable for that particular hotel - "A lot of specific technology is involved (p.c., Hotel A 2015, p. 5)." Systems are often designed and produced by large foreign companies that do not have a thorough understanding of local characteristics. Costly production and installation mistakes can be avoided by meeting with local contractors about property specifics for renewable energy installation before a system is purchased (p.c., Hotel B 2015, p. 13).

Solar energy is a terrific option for SIDs. Well-designed and properly installed systems are robust and require little to no maintenance (p.c., Hotel B 2015, p. 5). However, misinformation about technology specifics can discourage adoption. For example, in the BVI where outdated political legislation hinders the installment of renewables in areas served by the energy monopoly, government representatives claimed on local media that solar energy was unsafe as linesmen could be electrocuted through backfeed from rooftop panels (p.c., Hotel B 2015, p. 32). This is however, not the case. Solar energy has been safely and widely implemented around the world for many years (p.c., Hotel B 2015, p. 32). Nevertheless, when industry professionals publicly distribute such misinformation on small islands it can quickly spread and diminish public support for hotel initiatives. Wind is another popular option for SIDHs. It is however, more viable for hotels with higher investment opportunities and larger energy needs (p.c., Hotel B 2015). As wind-energy equipment consists of moving parts, the cost of maintenance can be significantly higher than with solar (p.c., Hotel B 2015). Therefore, before choosing this option, hoteliers should make sure that there is an availability of skilled workers in the area who know the intricacies of such equipment and how it is to be maintained.

No matter which renewable energy option is chosen, it is important that hotels conduct an environmental assessment of proposed projects and assess the best ways for minimal impact on the environment and guest experience. For example, with wind power a hotel can paint the turbine to match the landscape for aesthetic purposes but must take adequate measures to ensure that does not cause bird collisions (p.c., Hotel A 2015, p. 4). Just as discussed with solar, it is important that hoteliers and decision makers are aware of wind energy realities. There is a myth that the carbon footprint associated with the production of a wind turbine is higher than its offset emissions. Recent studies have confirmed that this is not the case (p.c., Hotel A 2015, p. 6). In fact, this has been calculated by Hotel A, where due to the amount of diesel being reduced through the installation of one 900 kw turbine, the production footprint should be offset in approximately one year (p.c., Hotel A 2015, p. 6). People must be educated on the realities of modern technologies so that barriers associated with misinformation can be avoided.

3.4 *Water Management*

An area directly related the ability of SIDHs to accommodate guests is water management. Many parts of the world are already feeling freshwater stress associated with climate change (Jones et al. 2014, p. 9; p.c., Hotel B 2015, p. E. 19). Many islands do not currently have a water management policy for hotels (p.c., Government Representative 2015, p. 2). However, even without governmental legislation, in order for hotels to maintain operations, water management is critical. "If we run out of water we cannot run our business. We cannot make money (p.c., Hotel B 2015, p. 19)." As a first step to dealing with water limitations hotels must monitor and measure consumption. It is also important to identify the most realistic and sustainable sources of fresh water. Aside from government purchased water, which due to transportation cost and disturbances is not a viable option for many SIDHs, sources of fresh water include rainwater collection and storage in cisterns as well as reverse osmosis. Many SIDHs use a combination of both. Without extensive guttering opportunities, the collection of rainwater will likely not be adequate to supply all water needs. However, it should still be integrated as much as possible throughout properties. Reverse osmosis is a great option for hotels and can be made sustainable by running the system on renewable energies. Three of the four hotels interviewed relied on desalination as the primary source of fresh water (p.c., Hotels A, B and C 2015). The fourth hotel relied on a combination of cistern collection and water purchased from the local water supplier. The reason being that this hotel was located in an area where the use of renewable energies is made difficult through governmental legislation with regards to the energy monopoly - making it unsustainable and too expensive to run a reverse osmosis system on fossil fuel based energy sources (p.c., Hotel D 2013). In order to fully combat water scarcity hotels must not only develop innovative sources of fresh water collection, but also include water conservation and reduction initiatives.

Hotels are required to offer their guests clean drinking water and plastic water bottles are not a sustainable option. Misconceptions about the feasibility of creating a clean source of drinking water leads to challenges involved with convincing staff and guests to choose tap water over bottled. Many residents come from Caribbean communities that lack clean sources of municipal drinking water and therefore, when a hotel offers filtered water (often of higher quality than the locally available bottled water) people hesitate to drink it (p.c., Hotel B 2015, p. 25). "Many people are falsely informed about issues such as bottled water. They will not drink tap water because they think it is dirtier than bottled. The truth is that bottled water is more loosely regulated than tap." Misconceptions like this lead to people leading less sustainable lifestyles for no other reason than lack of awareness (p.c., Hotel A 2015, p. 13). "Its extremely easy to make drinking water. People think it is hard. It is not. It's one of the easiest things (p.c., Hotel B 2015, p. 25)." Furthermore, SIHs should view water management as a top priority and form partnerships with each other, local businesses, government and the community to take advantage of local knowledge and organizations tackling the issue (p.c., Water Management Expert 2015).

3.5 Waste Management

Waste streams are critical for hotels on small islands to manage as they do not have the WM capacity of mainland areas. Therefore, it can become a huge issue and degrade both the aesthetic appeal and ecosystem functions of the environment, which the hotel relies on as its core product. Key areas of WM for hotels on SIDs are plastic, glass, sewage, compostables, and toxins. Improving WM practices and implementing a plan will significantly contribute to a safe and sustainable environment. Utilizing available resources optimally improves human health, creates jobs and can harness creative potential to transform current waste challenges into opportunities for betterment (IRF 2012, p. 176).

Similar as discussed above with energy, small islands commonly have weak or out-dated pollution legislation (IRF 2012, p. 40) and hotels are therefore, not legally required to properly manage waste. If pollution legislation is not clearly defined, significant room for individual interpretation by businesses can lead to loose WM practices. Furthermore, legislation that is not adequately enforced, which commonly occurs on SIDs, means that hotels must be self-motivated with WM and specifically train staff and inform guests about the importance of proper disposal and waste reduction both when at the hotel and elsewhere. Aside from weak environmental legislation, barriers that prevent SIDHs from participating in local WM programs (assuming they exists) are lack of doorstep collection as well as the time and training cost of separation (Radwan et al. 2009, p. 187). Hotels could significantly reduce separation costs by eliminating waste throughout the supply chain. Lifecycle assessments of products should be conducted before purchasing and products chosen, not only based on price, but also quality and minimal packaging consisting of materials that can be easily disposed of within the local system (Radwan et al. 2009, p. 187). This should be tailored to the specific island. For example, it does not help to order products consisting of recyclable plastics if there is no way of processing them. In such cases, it is much more valuable to focus on biodegradable materials and overall reduction (Radwan et al. 2009, p. 187).

Waste management is much more viable where economies of scale exist (p.c., Hotel B 2015, p. 24). Therefore, hotels should form partnerships with each other, other local businesses, NGO's, government and community members to benefit from larger streams. If WM companies could be guaranteed that the majority of businesses on the island would use their service, they will be more likely to invest in high quality recycling equipment (p.c., WM Expert 2015). Waste management is not usually identified as a priority issue in island development planning; at the same time, most small developing islands lack sufficient institutional resources (human, technical and financial) to handle increasing and more complex streams. In addition, the finite land area limits disposal options, thus increasing the urgency to address WM issues (IRF 2012, p. 166). Furthermore, the responsibilities of hotel staff rarely include the entire waste disposal cycle. The environmental impacts of poor WM practices include, soil, water, and air pollution as well as all associated environmental degradation and health concerns (IRF 2012, p. 173). However, many small islands lack the technical and financial resources and proximity to associated industries to properly treat and recycle materials (IRF 2012, p. 172). Dependency on external markets and imports skew economies of scale against WM opportunities for SIDs, which leads to high managing costs for relatively small waste quantities (IRF 2012, p. 166). Within these constraints however lies the opportunity to develop island specific innovative solutions for the management of increasing volumes (IRF 2012, p. 116).

Even with the associated complexity of WM on islands, there are examples where more sophisticated WM operations in which all waste streams are either fully processed on-island or are processed for recycling before leaving the island (p.c., WM Expert 2015, p. 1). Hotels or their contracted WM companies, should identify nearby processing and recycling hubs and build partnerships in those areas. Furthermore, hotels and governments should join to provide a grant system for local businesses or NGO's wanting to invest in WM operations as the initial investment in equipment can make projects unfeasible for local companies (p.c., WM Expert 2015, p. 3).

Plastics have an immense impact on marine ecosystems (Eriksen et al. 2014). The elimination of plastics and plastic packaging is one of the biggest and most necessary challenges for hotels (p.c., Hotel A 2015, p. 9). Tourists are often unaware of the delicate ecological balance associated with SIDs (IRF 2012, p. 40). Therefore, it is the responsibility of hotels, suppliers and governments to eliminate plastic packaging from being imported in the first place. Plastic reduction should be at the forefront of all purchasing decisions made in hotels. Using glass could be a more sustainable option for SIDs as it is even cheaper than plastic to produce and does not have such a lasting impact on ecosystems. Unfortunately, it is much heavier, which leads to higher transportation costs and the increased energy associated (p.c., Hotel B 2015, p. 29). Investment in a refined glass making process resulting in a lighter product would be greatly beneficial for SIDs (p.c., Hotel B 2015, p. 30). Aside from this, there is the additional issue of glass not being an option for SIDHs located on beaches as glass can break and be dangerous for visitors. There are many options for the processing of used glass on SIDs to avoid landfill. Some provide better solutions than others. Aside from classic approaches, more innovative methods of glass processing exist (Green VI 2015, p. 41). Such solutions should be shared by SIDHs globally and the most appropriate methods integrated at the local level.

Biodegradable material is perhaps the easiest for hotels on SIDs to manage (IRF 2012, p. 176). Hotels should focus on stocking compostable products with no packaging, such as local fruits and vegetables as well as choose biodegradable packaging over conventional plastic. Composting is an effective and simple WM strategy. The burning or burying of organic waste renders this valuable product useless (IRF 2012, p. 176). Organic waste composes very quickly in hot island climates and composting utilizes the natural peptides derived from organic material thus, saving money and protecting soil quality by reducing reliance on fertilizer imports (Purkis and Miller 2012, p. 14). Furthermore, organic waste represents a large percentage of a hotel's waste stream and therefore, composting it can significantly reduce dioxin emissions (IRF 2012, p. 176). Hotels should collect all organic material from restaurants and grounds for composting (p.c., Hotel B 2015, p. 14). If an animal shelter, humane society, or farmers are available in the area, food waste can be donated (p.c., Hotel B 2015, p. 23).

Proper sewage management is critical for hotels on SIDs. Untreated sewage and wastewater disposed into costal water poses health risks to visitors and residents. Increased visitation presents the need for updated sewage disposal practices to prevent disease-producing bacteria build-up in near shore environments and ecosystems (IRF 2012, p. 178). Hotels should encourage local governments to develop a national policy for wastewater management that forbids the disposal of untreated sewage in marine and terrestrial environments (CANARI 2012, p. 11). On small islands that lack either central or private sewage treatment facilities, sewage is dealt with via septic tanks, sump-holes, field beds, or soak-aways. The most common of which is the septic tank. Septic systems pose many challenges for hoteliers on islands. Systems commonly malfunction due to variations in weather and use. Overflow can occur during heavy rain events and transport effluent into coastal waters leading to the deterioration of water quality (IRF 2012, p. 178). Therefore, hotels should invest in sewage treatment systems. When calculating the cost of septic tank pumping, overflow events, and associated ecosystem degradation, investment in a properly designed sewage system will likely quickly pay for itself (p.c., Hotel B 2015, p. 9). Fluctuating guest numbers due to seasonality or the hosting of large events such as conferences and weddings, poses a serious problems for sewage treatment at SIDHs (p.c., Hotel A 2015, p. 8). Hotels should research and develop innovative strategies to deal with challenges involved with the carrying capacity of sewage storage and treatment systems. This can help to better deal with obstacles such as weather inconsistencies and guest fluctuation. Sewage released into the local environment is not only dangerous to ecology but also poses health threats for guests and is therefore, necessary for hotels to properly manage.

The management of toxic waste can be challenging for hotels on SIDs as many will not have access to appropriate disposal systems. Sources of toxins include items such as, batteries, aerosols, landscaping products, cleaning products, and cigarette butts. Landscaping and gardening is another area where hotels on SIDs should focus on toxin reduction. No toxins should be used in these areas. "There are natural ways of managing landscapes which work just as well (p.c., Hotel A 2015, p. 11)." Hotels should talk to local farmers about what natural methods work best in their region (p.c., Hotel B 2015, p. 20). The above discussion offers a brief glimpse into the WM challenges associated with hotels on SIDs and offers some solutions. However, as many more challenges and solutions exist in this area, hoteliers are encouraged to review hotel specific sustainability guidelines and partner with local stakeholders to develop innovative solutions.

3.6 *Wildlife Conservation Management*

The seclusion of SIDs, means that ecosystems have been protected by remoteness (Boniface and Cooper 2009, p. 46) and have developed with little to no human interaction until the development of tourism, therefore, animals with no natural predators are not afraid of humans, leaving them more susceptible to disturbances (Boniface and Cooper 2009, p. 49). As ecotourism continues to grow in popularity, public interest in conservation initiatives on SIDs offers an opportunity to provide support for conservation, but on the other hand increases the popularity of fragile ecosystems posing additional threats (Boniface and Cooper 2009, p. 46). Strategies for the minimization of anthropogenic threat sources, such as increased development pressure, oil spills, and ship groundings must be addressed by hoteliers on SIDs (Gardner et al. 2008, p. 83). Local governments should aid businesses in the protection of ecosystems against anthropogenic threats, however this is often not the case. Therefore, the capability of institutions working on ecosystem protection on small islands may be severely limited by inadequate financing and limited budgetary support (Gardner et al. 2008, p. 83).

Due to the lack of governmental support, private sector institutions, firms, and individuals will be encouraged to play an increasing role in protected area management (Gardner et al. 2008, p. 23). Hotels must learn to manage and monitor their surrounding environment and implement biodiversity management guidelines (UNWTO 1998, p. 698). Wildlife conservation can become an integrated part of the hotel experience (De Silva 2001). An integrated wildlife conservation program is easier to maintain if staff in seemingly unrelated areas are encouraged to participate in conservation projects. Furthermore, a conservation foundation should be set up so that guests can easily make donations to support projects (p.c., Hotel A 2015, p. 12). If environmental capacity is not adequately measured and its limits respected, damage to ecosystems can occur such as nesting disturbance and erosion of paths to highly visited areas (Boniface and Cooper 2009, p. 49). As most islands are visited because of their environmental resources, disturbances intern affect the value of hotel businesses. A lack of capacity with regards to the natural environment is not the only area of insufficiency—it also commonly exists within the organization. Inadequate management capacity for conservation affects the monitoring, planning, enforcement, and maintenance of institutions resulting in increases in species loss (Gardner et al. 2008, p. 83).

Reefs are a common feature and tourist attraction of SIDs. Around the world reefs are facing enormous degradation from development, pollution and climate change. This is predicted to increase (Benckendorff and Lund-Durlacher 2013, p. 57). Therefore, hotels need to act now as the longer they wait the more difficult and costly reefs will be to protect. As climate change progresses the long-term outlook of reef health decreases. Diminishing water quality from catchments and runoff leads to costal development associated habitat loss (Benckendorff and Lund-Durlacher 2013, p. 58). The documentation and assessment of routinely monitored reef changes helps to maintain the delicate balance of health components at various trophic levels (IRF 2012, p. 136). Hotels located near reefs can help with the monitoring and recording of ecosystems by installing underwater cameras that live stream and record reef activity—this not only has the ability to provide valuable data to research organizations, but also educates guests on conservation (p.c., Hotel B 2015, p. 19).

Turtles represent a highlight for tourists visiting many SIDs. Guests may interfere with the natural ecology of beaches diminishing habitats and nesting success. Also indirectly, more demand for energy and supplies to support hotel development and operations increase motor activity in marine environments (IRF 2012, p. 146) and the likelihood of oil spills. Furthermore, hotels often want manicured beaches free of vegetation and wave-deposited organic material, which also diminishes the success of turtle nesting (IRF 2012, p. 139). Potential areas of turtle conservation for private institutions (such as hotels) to get involved in are financing, monitoring, and surveillance (Gardner et al. 2008, p. 23). Hotels should identify whether or not their beach is a turtle nesting area and if so, take adequate protection measures to preserve habitats, such as the reduction of light pollution. The Turtle Tool Kit offers valuable information on steps that can be taken to identify and protect turtle nesting sites (Varela-Acevedo et al. 2009, p. 11). Hotels can contact a local conservation group, the fisheries office, or responsible party for location specific conservation protocols (Varela-Acevedo et al. 2009, p. 9).

Ecosystem survival on SIDs is negatively affected by numerous serious problems associated with environmental interference by humans (Boniface and Cooper 2009, p. 46). Attempts to implement quarantine systems have commonly proven to fail, sometimes resulting in the amount of introduced flora and fauna species outnumbering native ones (Boniface and Cooper 2009, p. 46). The impact of non-native flora and fauna on fragile island ecosystems is well documented and of great concern. Small tropical islands worldwide seem especially vulnerable. The invaders have been responsible for the demise of numerous endemic plants and animals. Discussion on invasives is primarily terrestrial in focus, and there is usually little mention of non-native marine organisms (IRF 2012, p. 136) although numerous examples exist.

3.7 Food and Beverage

Hotel businesses are in constant need of food and beverage supplies for restaurants and bars. Remoteness poses many challenges (p.c., Hotel A 2015, p. 10). A steady supply of high quality ingredients can be difficult to acquire and choice is often severely limited and imports expensive. Hoteliers must be committed to the sustainable sourcing of food products and demonstrate this the establishment of supply relationships with local farmers (Jones et al. 2014, p. 10).

Example initiatives that can be taken to ensure the sustainability of food products are:

- Acquire certification by a Sustainable Restaurant Association.
- Have chefs and suppliers complete a sustainable sourcing program.
- Make marine health a priority by avoiding products that remove fish, lobster and other ocean delicacies from fragile ecosystems.
- Offer vegan and vegetarian options.
- Encourage guests to advise the hotel of dietary requirements upon booking to avoid excess food stocks.
- Purchase local and organic labelled products when available (p.c., Hotel A 2015).

As stated above, the availability of local and organic products can be a challenge for SIDHs (Purkis and Miller 2012, p. 14). However, suppliers want the steady business associated with hotels that require a constant supply of repeat products. Therefore, giving them an incentive to stock the products in demand by local hotels. Hotels should not just wait for suppliers to stock the products they want, but rather take active steps to make them available. This can be achieved by telling suppliers exactly what they are looking for and asking them to make it available to get the hotel's business. Hotels should as much as possible integrate flexibility into menus to allow for the purchasing of on-island agricultural produce from local farmers (p.c., Hotel A 2015, p. 10). If sustainability is to be a priority, barriers must be overcome by a change in mindset and the prioritizing of sustainability over small discrepancies such as menu standardization. Local initiatives should always be supported instead of outsourcing as it encourages the continuality of such projects and the development of new supply sources. In addition to the support of local agriculture, hotels should focus on on-site production of food—either for guests, staff or to support wildlife conservation projects. "The island needs to become more self-sufficient. If fruit trees were planted in a few years food (for the animals) would no longer have to be purchased. It would also have a positive impact on guest perceptions (p.c., Hotel A 2015, p. 11)."—stated the Wildlife and Conservation Manager of Hotel A.

Composting can help to increase soil quality for on-site agriculture (p.c., Hotel B 2015, p. 14). It will likely be difficult to fully supply the restaurant with on-site produce, which is why (as stated above), it should be supplemented by local partnerships (p.c., Hotel B 2015, p. 14). Examples of challenges for on-site cultivation on many SIDs include the existence of high saline levels in soil and air as well as making sure that gardens do not become a breeding ground for mosquitoes (p.c., Hotel A 2015, p. 10). Which agricultural products will grow best on an island is area specific and hotels should research the best solutions for their property (p.c., Hotel B 2015, p. 13). Furthermore, hotels should review which products they consume the most of and try to localise them as much as possible to cut down on cost, waste, energy, and pollution (emissions from ships and increased possibility of oil spills). If SIDHs require external financing or design options for sustainable solutions (p.c., Hotel C 2015, p. 8), they can take advantage of global community based funding and innovation opportunities such as crowd funding and open-innovation platforms (e.g. innonatives.com).

4 Concluding Remarks

Tourism represents the world's largest service industry (Lew 2008). Like a global eco-system, the tourism industry is integrated with an enormous variety of international economic activities. The hospitality sector is a key component. Although it is difficult to calculate an exact number, it has been estimated that approximately 500,000 hotels exist worldwide (Spotted by Locals 2012). Together, these hotels can have an outstanding influence on environmental, social and economic systems. It is extremely important that hospitality organizations be developed and managed in a sustainable manner. Sustainability within a company can be an extremely complex and a difficult metric to measure requiring a highly developed and integrated management system - especially on SIDs.

It should be noted that (as is the case in many destinations) in the BVI, although there are sustainability goals and measures within the government, the majority of successful projects have been initiated by private businesses, NGO's, and locals who receive little to no support. At times they are even hindered by out-dated policies and short term profits (p.c., WM Experts A and B, Hotels B and D, and Agricultural Expert 2013–2015). Many of the tourism developments, challenges, and opportunities discussed in this paper are relevant for SIDs around the world. The researcher strongly encourages further research in the overview of this topic as well as in-depth research into categories identified as critical for hotel sustainability on small islands. Sustainability challenges and opportunities must be thoroughly discussed in academia and practice. Action needs to take place now if hotels will be able to cope with challenges brought upon by external factors such as climate change, environmental degradation, and human development. This should be a global effort. Sustainable hospitality management has the ability to create positions for skilled workers, thus increasing employment and in turn creating a more secure tourism industry. There exists an abundance of innovative research and business opportunities to be developed within this area and more investment should be placed on the value of such discussions.

5 Limitations

Time was the most significant limitation for this research. Both sustainable tourism management and SHM for small islands are huge topics of global relevance. There are few studies that focus on the overall concept of how the two are connected from a strategic hotel management perspective. In order to develop a holistic approach to this topic, numerous hotels on small islands around the world would have to be researched. This was not possible due to the scope of this project. Focus was therefore placed on researching the Caribbean, primarily the BVI. However, if a longer research period were possible, SHM in this region could have also been much more comprehensively reviewed. Many additional experts and locals in the field were willing to participate in interviews or help provide documentation on the topic. Much of this was however not possible to conduct and integrate due to constraints. Furthermore, there are likely differences between how hotels should be managed on islands that experience colder climates in comparison to warmer ones. Differences in other geographic and cultural features should also be considered. This study primarily focused on "Sun and Sand" tourist destinations with the majority of information coming from Caribbean studies. Other regions, which were mentioned however, less thoroughly researched, include the South Pacific (Galapagos Islands) and the Indian Ocean (Mauritius). Additionally, due to the type of hotel development in the region where primary research was conducted, many of the hotels interviewed were high end and/or on the smaller side. Further research should be conducted for other varieties of hotels (large, budget, all inclusive, etc.). Charter boats and cruise ship accommodation was not included in this research which is a significant area of interest for SIDs.

An abundance of literature exists on sustainable hospitality management, sustainable island management and sustainable tourism management. However, there is a lack of published information specifically combining all three. This created both an opportunity and constraint during the research process as combining the topic is a large task requiring more time than allotted. Therefore, due to time and scope limitations there were many areas, which were not covered including the social dimension of sustainability. The goal of SHM research for SIDHs should be to assess individual situations with the base of a theoretical framework to come up with innovative solutions to specific challenges. This topic is so broad that a number of PhDs could be written on the specifics of each category and sub category featuring different islands around the world.

Appendix

List of Personal Communications (Location British Virgin Islands)

Hotel A

- Deputy Director, Chief Engineer, and Wildlife and Conservation Manager
- 15-room five-star resort
- Duration: Approximately 3 h
- Location: on-site
- Date: February, 2015

Hotel B

- General Managers (one GM also the Chief Engineer)
- 9-room resort
- Duration: Approximately 4 h
- Location: on-site
- Date: February, 2015

Hotel C

- Food and Beverage Manager
- 100-room five-star resort
- Total duration: Approximately 30 min
- Location: on-site
- Date: February, 2015

Hotel D

- Both Owners/Managers
- 11-room resort
- Duration: Approximately 2 h
- Location: On-site
- Date: January, 2013

Agriculture Expert

- Local farmer and sales partner
- Long-term community members (over 20 years)
- Duration: Approximately 40 min
- Location: On-site
- Date: February, 2015

Waste Management Expert

- CEO and Founder of recycling business
- Over 20 years in the community
- Duration: Approximately 40 min
- Location: OCTA Ministerial Conference Exhibition
- Date: February, 2015

Sustainability and Waste Management Expert
- Executive Director of a BVI NGO
- Over 15 years in the community
- Duration: Approximately 25 min
- Location: On-site
- Date: February, 2015

Government Representative (Environment)
- Senior Officer of the BVI Tourist Board
- Duration: Approximately 15 min
- Location: BVI restaurant
- Date: February, 2015

Water Management Expert
- CEO of a BVI water purification and sewage treatment company
- Duration: Approximately 1 h
- Location: On-site
- Date: February, 2015

References

Benckendorff, P., & Lund-Durlacher, D. (2013). *International cases in sustainable travel & tourism* (pp. 57–113). Oxford: Goodfellow Publishers.

Boniface, B., & Cooper, C. (2009). *Worldwide destinations casebook. The geography of travel and tourism* (2nd ed.pp. 45–251). New York: Elsevier.

Caribbean Natural Resources Institute (CANARI). (2012). Environmental mainstreaming in the British Virgin Islands - final project report. Greening the economy - towards sustainable development for the BVI.

De Silva, R. (2001). Kandalama Hotel. Retrieved April 4th, from http://ecolodgical.yourhomeplanet.com/pdfs/case_study_kandalama.pdf

Eriksen, M., Lebreton, L. C. M., Carson, H. S., Thiel, M., Moore, C. J., Borerro, J. C., et al. (2014). Plastic pollution in the world's oceans: More than 5 trillion plastic pieces weighing over 250,000 tons afloat at sea. *PLoS One, 9*(12), e111913. doi:10.1371/journal.pone.0111913.

Gardner, L., Smith Abbott, J., & Woodfield-Pascoe, N. (2008, January). *British Virgin Islands protected areas system plan 2007–2017*. Tortola: BVI National Parks Trust.

Green VI. (2015). *Green VI's glass studio*. Retrieved May 1, from http://www.greenvi.org/Green-VI-s-Glass-Studio.aspx

International Tourism Partnership (ITP). (2014). Environmental management for hotels: Energy. *The industry guide to sustainable operations*. Retrieved May 1, from http://www.greenhotelier.org/category/our-manuals/environmental-management-for-hotels/

Island Resources Foundation. (2012). *An environmental profile of the Island of Virgin Gorda, British Virgin Islands, including Eustatia, Mosquito, Necker, Prickly Pear, Saba Rock, The Dog Islands, Broken Jerusalem, Fallen Jerusalem, and Round Rock* (p. 255). Island Resources Foundation, Tortola, British Virgin Islands and Washington, DC.

Jones, P., Hillier, D., & Comfort, D. (2014). Sustainability in the global hospitality industry. *International Journal of Contemporary Hospitality Management, 26*(1), 5–17.

Lew, A. (2008). *Tourism is NOT the world's largest industry – so stop saying it is!*. Retrieved March 1 from http://tourismplace.blogspot.ca/2008/04/tourism-is-not-worlds-largest-industry.html

Prayag, G., Dookhony-Ramphul, K., & Maryeven, M. (2010). Hotel development and tourism impacts in Mauritius: Hoteliers' perspectives on sustainable tourism. *Development Southern Africa, 27* (5 (Special Issue: Tourism impacts: Lessons for policy, programmes and projects)), 697–712.

Purkis, J., & Miller, K. (2012). British Virgin Islands sustainable capacity building program summary report. *The Natural Step*.

Radwan, H. R. I., Jones, E., & Minoli, D. (2009, March). Managing solid waste in small hotels. *Journal of Sustainable Tourism, 18*(2), 175–190.

Rogers, T., Chmutina, K., & Moseley, L. L. (2012). The potential of PV installations in SIDS – an example in the island of Barbados. *Management of Environmental Quality: An International Journal, 23*(3), 284–290.

Spotted by Locals. (2012). *How many hotels are there in the world?*. Retrieved March 1 from http://www.spottedbylocals.com/blog/booking-com-and-the-race-towards-offering-all-hotels/

Summit Secretariat (2013). Summary of CCI summit outcomes – outcomes of the Caribbean summit of political and business leaders to launch the second phase of the Caribbean Challenge Initiative (CCI). *British Virgin Islands.*

UNESCO. (2010). *Adapting to climate change and educating for sustainable development. Sandwatch.* Paris.

United Nations World Tourism Organization (UNWTO). (1998). *Guide for local authorities on developing sustainable tourism.* Madrid: UNWTO.

Varela-Acevedo, E., Eckert, K. L., Eckert, S. A., Cambers, G., & Horrocks, J. A. (2009). Examining the effects of changing coastline processes on Hawksbill Sea Turtle (*Eretmochelys imbricata*) Nesting Habitat. *Sea Turtle Nesting Beach Characterization Manual* (pp. 46–97). North Carolina, USA.

Accounting for Sustainability: The Case Study of Petrobras

Rute Abreu, Fátima David, Liliane Segura, Henrique Formigoni, and Flávio Mantovani

1 Introduction

The quality of information reported by the public companies has been highly discussed along the last decade. The center of this discussion is the implementation of International Financial Report Standards (IFRS) in many countries around the world, searching for reports that could be read for the investors and bring enough information to decide rationally the resource allocation inside the financial system.

However, it has been known that this implementation can be very different depending on the country and the legal system that this company is inserted. Also it also has been discussed that the inside governance system, even clear politics has been influenced by the principal investor and, once that the main shares are concentrated in only one stockholder, the influence in accounting policies could damage the quality of accounting information.

The purpose of this paper is to demonstrate the adoption of the appropriate accounting framework supported on the International Accounting Standards (IASB 2015) and the influence of International Auditing Standards (IFAC 2015) on the promotion of a corporate governance strategy approved by companies to increase the transparency of the information. Furthermore, the authors will be focus on the professional skepticism in relation of the fairness of assertion on the financial statements and the management responsibility for global interest in corporate responsibility and sustainability.

R. Abreu (✉) • F. David
Guarda Polytechnic Institute and UDI-IPG and CICF-IPCA, Guarda, Portugal
e-mail: ra@ipg.pt; sdavid@ipg.pt

L. Segura • H. Formigoni • F. Mantovani
Universidade Presbiteriana Mackenzie and Mackpesquisa and CCSA-UPM, São Paulo, Brazil
e-mail: lilianecristina.segura@mackenzie.br; henrique.formigoni@mackenzie.br; flavio.mantovani@mackenzie.br

H. Lu et al. (eds.), *Building New Bridges Between Business and Society*, CSR, Sustainability, Ethics & Governance, DOI 10.1007/978-3-319-63561-3_8

The main contribution of this research is to explore the accounting communication with the society made by Petrobras as the largest company in Latin America and one of the biggest integrated energy companies in the world, with a strategic role in oil geopolitics. Then, the authors aim to understand the relevance of Petrobras activity through the accounting processes and the financial report that it is strongly regulated as a fundamental area of the financial markets. As the future development, the authors aim to widespread the accounting and financial report as weapon to combat mal-functioning of the financial markets and help several stakeholders.

2 Theory

2.1 IFRS and Quality of Accounting Information

Increasing the current countries that have adopted the international accounting standards (IFRS), the Brazilian accounting was inserted in this context through the enactment of Law 11.638-07 (Government of Brazil, Ministry of Finance 2007). However, in 2005 it had been created the Accounting Pronouncements Committee (CPC) through the Resolution 1055-05 of the (CFC) Federal Accounting Council in order to study, prepare and issue the accounting statements (CFC 2005, article 3):

> (. . .) Technical Statements on accounting procedures, disclosure and other matters of such nature, to release standards by the Brazilian agency, aimed at centralization and standardization of the production process, taking into account the convergence of the Brazilian accounting to the international standards.

In Brazil, the adoption of international standards was carried out in two different times. The first hybrid, covering only those set of financial statements released to date and since 2010 with the full adoption. The statements set by the International Accounting Standards Board (IASB) are primarily designed for the countries, which have developed capital market. The application of IFRS in countries with economies in development and transition is more difficult, due to the lack of structure to monitor the decisions taken about the publication and disclosure of financial statements (Peng et al. 2008).

In parallel to the efforts of dissemination and training, Baptista (2009) warned of the need for discussions on the possible effects of a major conceptual changes brought by the new law, which is the biggest discretion given to preparers of financial statements. According to the author, the greatest power of judgment awarded by international standards can contribute to the design of a more favorable environment for the practice of manipulation of accounting results.

The harmonization process, even if mandatory, can lose strength if companies realize that the consequences in case of non-compliance with the new standards are not severe (Tay and Parker 1990). Maia (2011) states that the accounting science is shaped and developed under the direct influence of numerous environmental factors that surround it, thus, it is not possible to separate the creation of accounting practices of the legal, political and taxation systems and the economic, cultural and historical factors of a country.

Murcia et al. (2008) explain that the largest companies as well as those with higher profitability tend, in the first case, to disclose more information to enhance their corporate reputation and second, to show they perform better than its competitors that in its turn could generate a reduction on the fundraising costs.

In order to ensure more reliability to the users of accounting, the financial statements are audited by independent auditors in order to give an opinion to clarify whether they adequately reflect the financial position of the company. Thus, the auditor should use criteria that produce elements to ensure the effectiveness of the figures showed in the financial statements, and should perform procedures to enable it to express an opinion whether the financial statements fairly represent the financial position of the company.

De Angelo (1981) states that audit quality is influenced by the size of the external audit firm, as the larger accounting firms, as so called big four, tend to lose more with some inconsistency in their opinions than a smaller auditing firm . In addition, larger audit firms tend to have greater independence from its customers, reporting errors more frequently and pressing their clients to have a high level of compliance with legal and statutory requirements.

However, Uwuigbe (2011) states that the empirical evidence on the relationship between the size of the audit firm and the amount of information disclosed by the companies are emphasizing controversial studies that indicated a positive relationship between audit firms and voluntary disclosure (Hossain et al. 1994; Ng and Koh 1993) and others found no such relationship (Malone et al. 1993; Mohamed and Janggu 2006).

Transparency, clarity and comprehensibility of the information contained in the financial statements are inherent to good corporate governance practices, which in turn, lead to a lower risk perceived by the investor. Thus, investors may demand a lower return on investment enabling a reduction in the cost of capital of the companies (Antunes et al. 2007).

Chen et al. (2004) state the need of a set of mechanisms for the protection of investors against expropriation by managers. Regarding corporate governance, they claim that the best corporate governance reduces:

1. The relationship between the degree of expropriation and market conditions,
2. The cost of equity, by reducing the cost of monitoring by investors, and
3. Information asymmetry between investors and managers.

Thus, the level of corporate governance can influence the degree of minimal disclosure of business and should contribute to increasing the usefulness and quality of accounting information.

Concerning to Earnings Management, Paulo and Martins (2007) found evidence that the accounting reports of Brazilian companies are less conservative and have greater earnings management level than those of US companies, suggesting therefore that there are differences in the quality of accounting information originated by the economic environment in which business operates.

On the disclosure of information in the financial statements, Uwuigbe (2011) advises that there are several studies measuring the disclosure of financial and non-financial citing as examples the studies of Ahmed and Nicholls (1994), Buzby (1974), Cooke (1989), Hossain (2000), Inchausti (1997), Kahl and Belkaoui (1981), Singhvi and Desai (1971), Wallace (1987), and Wallace and Naser (1995). Although there have been many of the studies in this matter, the quality of accounting information is still a topic that needs more thorough and robust research.

2.2 Adoption of IFRS Effects in Brazil

Moraes et al. (2014) analysed the impact of the accounting convergence process with international standards in the net income of the legal entities in Brazil. Data were collected from Statements of Economic-Tax Information of Legal Entities (DIPJ) and covered the years 2008–2011 by those companies that have opted for taxable income and were submitted to the differentiated economic-tax monitoring of the Federal Revenue of Brazil (RFB). The study totalled a sample of 8080 comments.

Data from the study variables were used to calculate the index Gray, in order to measure the change in book value under the two standards sets (BRGAAP × IFRS). The overall analysis showed that the book value under the influence of international standards increased in the 4 years studied. It was found that most companies of the sample had increased book value due to the effects of international standards. The impact of international standards was up early in the convergence and downward process in the final stage. Additionally, it was found that a small number of corporations had their profit modified to loss and vice versa due to the adoption of international accounting standards.

Domenico et al. (2014) found that the adoption of international accounting standards (full IFRS) by Brazilian companies brought statistically significant impact on the financial indicators of companies listed on BM&F Bovespa, referring to the financial statements for 2009, which were restated in 2010 according to full IFRS for comparison purposes. The results showed that the adoption of international accounting standards did not affect the financial indicators of the companies analysed. The impact observed was related only to immobilization of the permanent capital.

Costa et al. (2013) studied the existence of significant differences between the value of shareholders' equity, net income and return on equity (ROE) calculated in 2008 and 2009, transition period, under different accounting standards (IFRS and Brazilian GAAP). In addition, they sought to identify changes in accounting practices were more significant and frequent in this period.

The sample consisted of 14 public companies listed on the Bovespa belonging to the beverage segment, food and trade. The survey results showed: (a) a reduction of

27.6% in net income of 2008 and an increase of 1.4% in net income for 2009, both statistically significant (at 10%); (b) a reduction of 0.3% in equity 2008, an increase of 0.6% in the equity of 2008 and an increase of 5% in equity 2009, all not statistically significant; and (c) a change in the return on equity of 3 to 4% in 2008 and −132 to 26% in 2009.

Ponte et al. (2013) investigated the impacts of IFRS adoption in Brazil in representing the equity, economic and financial of the banks listed on BM&F Bovespa. The sample brought together 18 banks that reported its consolidated statements for 2010 in IFRS and Brazilian GAAP standards. Significant differences were observed in the Liquidity and Quality Loan Portfolio indicators, indicating that the financial statements prepared under IFRS signal less liquidity and quality of the loan portfolio compared to that evidenced by statements in BR GAAP. From the perspective of shareholders' equity, it was found that the financial statements in BR GAAP are more conservative than the statements under IFRS.

Martins and Paulo (2010) investigated the reflection of the adoption of IFRS on the performance indicators of Brazilian listed companies in order to identify if during the period investigated there was a reduction of differences between the calculated indicators from national accounting standards and international standards. They collected the financial statements of 13 companies listed on the BM&F Bovespa, prepared in accordance with Brazilian GAAP and the IFRS for the fiscal years 2007, 2008 and 2009, from which were analyzed seven performance indicators.

The results showed that the adoption of IFRS has been reflected in the performance analysis of companies through positive changes in indicators of financial dependence, indebtedness, return on assets and return on equity, and negative variations on immobilization indicators of permanent resources, general liquidity and current liquidity. However, the differences between the indicators calculated from two sets of standards have declined, mainly due to the increasing convergence of Brazilian accounting standards to international accounting standards.

2.3 *Quality of Accounting Information After the Adoption of IFRS*

Herculano and Moura (2015) analysed the influence of the level of concentration of capital on the quality of accounting information in Brazilian companies from various economic sectors listed on the BM&F Bovespa. The sample consisted of 222 companies, with data for the period 2010–2012. As a proxy to analyze the quality of accounting information, two features were used: the persistence of financial results and cash flow (Dechow and Schrand 2004) and timeliness of accounting information (Bushman et al. 2004).

Regarding the concentration of capital, it has been verified the total percentage of shares (common and preferred) held by the ownerships. The results showed greater persistence in financial results and cash flows and greater opportunity of accounting information in the companies with higher concentrations of capital, indicating that the concentration of capital can be an influential factor in the quality of accounting information.

Martins et al. (2014) analysed the consequences of the convergence with IFRS in the quality of accounting information and, more specifically, assessed whether these possible consequences behave differently for listed companies in different levels of corporate governance (DLCG), as opposed to companies in general. They used a sample of 119 companies listed in the periods defined as before, during and after the process of convergence to IFRS model. The metrics used information quality such as, relevance, timeliness and conservatism. The main evidence indicated that the conservatism of the studied companies decreased over the convergence process however, the degree of reduction of the companies listed in DLCG was lower than the others.

With regard to the timing, it was observed that the companies listed in DLCG presented more timely before the convergence process, however, the process of adoption of IFRS, the level of timeliness was not different between groups of companies. This suggests that the adoption of IFRS contributed positively in relation to the timing attribute for companies that are not listed in DLCG. It was also observed that the relevance of financial reporting has improved with the adoption of IFRS, and, before and during the partial adoption, information relating to the member companies to DLCG showed a higher level of importance than the others. However, with the IFRS full adoption, no evidences of distinction was found.

Santiago and Cavalcante (2014) found that the quality of financial information of publicly traded companies in the construction industry in Brazil underwent changes when the country was framed into the international convergence. The analysis of the effects of the adoption of IFRS was limited to two attributes on quality of accounting information, conservatism and persistence. The sample consisted of 23 companies in the sector and the period 2006–2012.

For the data studied was found evidences of conservatism in the accounting information, so that it did not identify structural break in the reporting period and that the data remained conservative throughout the period. It is noted, however, that separate analysis of data refers to inconclusive results with respect to conservatism. Regarding the persistence, it was identified its existence in the financial information for both the data set and when analyzing them individually.

Silva (2013) investigated the impact of the full adoption of IFRS in the quality of financial statements of Brazilian public companies in the period 2000–2011, using a non-probabilistic sample consisting of the member companies of the theoretical portfolio of IBrX-100, in addition to 30 companies that adopted IFRS in advance. The results indicated an increase in the quality of accounting information after full adoption of IFRS. Most disclosure provided by the full adoption of IFRS can be explained by the transaction log not previously accounted for. These findings imply that the accounting information in IFRS can be more useful to the various users of accounting, as analysts, which require more informative accounting numbers in order to assess risk and provide more accurate analysis and estimations.

Moura et al. (2013) analysed a set of family Brazilian companies from different economic sectors of the BM&F Bovespa, if those listed on different levels of governance, with smaller and more independent boards of directors had better quality of accounting information. The sample consisted of 96 family companies whose data is obtained for the period 2008–2010. In order to analyze the quality of information used four characteristics: persistence of results and cash flow (Dechow and Schrand 2004); conservatism (Ball and Shivakumar 2006); Opportunity (Bushman et al. 2004) and relevance (Ohlson 1995). The results showed that companies in different levels and with more independent advice had greater persistence, conservatism, opportunism and relevance of accounting information.

Maia et al. (2012) studied the factors affecting the level of minimum disclosure of Brazilian companies participating in the IBrX index in 2008 and 2009. Such list is composed by the one hundred most liquidity companies and concluded that the disclosure level is positively influenced by the variables: size of the external audit firm, internalization of the audited company (through ADR programs), corporate governance (market level) and level of debt.

Vieira (2010) studied the earnings management, the timely recognition of losses and the relevance of accounting information in order to identify if there was an increase in the quality of financial reporting. Used stratified samples from the full set of public companies in order to compare possible changes in the quality of financial information between the pre-and post-partial adoption of IFRS through analysis of six quality measures.

In all samples were found clues that suggest that there was an increase in the quality of financial information as measured by management through income smoothing. However, it was not possible to say when used to achieve the management goal. As for the timely recognition of losses, it was found conflicting results, some showing of quality improvement and others not.

For relevance of accounting information, evidence, in all samples, it is that there was an increase in the quality of the information. In general, the results are consistent with the literature, demonstrating that IFRS, even partially adopted, provide improved quality of financial information of publicly listed companies in Brazil.

3 Method

The methodological research is divided on two perspectives. The first perspective is theoretical supported on the literature review from accounting and auditing standards. Thus, the consequences of accounting options and complex inter-relationship with auditing generate knowledge that is not yet possible to quantify and to obtain complete information (Warren et al. 2014; Messier et al. 2008; Arens et al. 2012).

The second perspective is an exploratory, longitudinal and empirical analysis supported on public available sources (Yin 2012), such as the annual and inter-reports in the period 2008–2015 made by Petrobras (2009, 2010, 2011, 2012, 2013,

2014, 2015). Indeed, the disclosure of the financial statements is from 2008 to 2013 and, subsequently, changes made on 2014 and 2015 of the financial statements.

The research question allows seeing accounting differences between the periods before and after the "Operation Wash Jet" Research, due to the impacts of the re-publication of financial statements in order identify the quality of accounting and auditing information provided to external users during this period.

4 Data Analysis

4.1 History

Petrobras is a Brazilian company for oil and gas extraction, founded in 1953. In September 2013 received the award for Best Company in the Oil and Gas Sector by the yearbook Season Business 360. The company was chosen for the second consecutive year, which features the 250 best companies in various industries, analysing all dimensions of business. The largest company also was voted the seventh most valuable Brazilian brand in 2014. This is the seventh time that the company was among the top ten of the Interbrand ranking (Petrobras 2015)

Additionally, the company has been in the ranking to recognize the transparency of its financial statements of publicly traded companies with revenues exceeding $5 billion, according to Transparency Award 2014. In its 18th edition, it is a recognition of the transparency of companies in relation to information given in financial statements. The award is organized by the Association of Executives in Finance, Administration and Accounting (Anefac), in partnership with the Institute of Accounting Research Foundation, Actuarial and Financial (Fipecafi) and Serasa Experian (Petrobras 2015).

According to the data disclosed in the Financial Statements Notes to the 2014 Petrobras, published in 2015 (after a comprehensive review of accounting numbers), occurred in 2009 a so-called investigation "Operation Wash Jet" by the Brazilian Federal Police, seeking to ascertain practices money laundering by criminal organizations in several Brazilian states.

Also according to these Notes, throughout 2014, federal prosecutors focused part of their investigations into irregularities involving contractors and suppliers of Petrobras and discovered a larger scheme of overpayments, involving a large number of participants, including former—employees of the company. Based on the information available to the Company, the said scheme involved a group of 27 companies between 2004 and April 2012, were organized in cartel for contracts with Petrobras, imposing additional costs on these contracts and using these additional amounts to fund overpayments the political parties, elected officials or other political agents, contractors and suppliers employees, former employees of Petrobras and others involved in improper payments scheme. This scheme has been treated as "improper payments scheme" and these companies as "members of the cartel."

In the same period (2014), in Brazil, it was taking place in the Presidential Election. Little information was disclosed before the results of the election, which took place in November 2014. As above, there was evidence of irregularities in Petrobras, but no notes to investors, or a material announcement on the financial report of company (Petrobras 2015).

According to the company, when the company released its 2013 annual financial statements on 27 February 2014, also released its Reference Form in May 2014 and released its interim financial statements for the second quarter of 2014 on August 8, 2014. Until then there was no evidence available about the investigation of "Operation Wash Jet" that could have changed the outcome of the Company with respect to the fact that those statements fairly represent its financial position and the existence of undue payments scheme had not been made public.

As can be seen, since 2008, when international accounting standards were implemented, does not realize any movement on the loss of values in relation to projects. However, due to the oil value of the fall and the economic crisis of 2008, the company's shares fell and stabilized from 2008 to June 2014. This also indicates that the financial statements did not provide any indication that the company had some financial problem.

After the elections, they began to appear more evidence in the media that the problems of Petrobras were quite relevant. Because of the public pressure, the auditing firm responsible for auditing the 2014 financial statements refused to sign them. At this point, it is that investors began to see that there really was something wrong in the financial statements of the company.

The delay in publication of Petrobras's financial statements caused several lawsuits filed by the SEC, requiring transparency of presented statements. The interim statements for the third quarter 2014 were only published after great pressure and in April 2015. Then were also published the statements of year 2014.

First, it is important to assess how Petrobras financial statements before the scandal of Operation Wash Jet were disclosed. An analysis of statements, shown in Table 1 was made. Figure 1 shows the change in the price of the company's shares during the period 2000–2014.

The delay in the dissemination of information caused a lot of distrust among investors. The Brazilian Securities Commission and the SEC also opened investigation against Petrobras to investigate the irregularities in the company. At the present time (June/2015), investigations are still ongoing, since it does not follow exactly what happened to the company, except that a lot of information was concealed in the financial statements.

What could be identified between the two periods is that, despite all the scandals, the accounting numbers fail to identify the actual losses incurred in the company. It was made a loss of impairment of assets in the magnitude of US$44 billion as informs the note 10.1 of the financial report of company (Petrobras 2015).

Low by formation of impairment in the amount of RM $44,537, which especially reflects the assessment of recoverability of assets related ace refining, exploration and production and petrochemical, resulting from changes in the Company's business of setting. For more information, see Note 14 of the financial report of

Table 1 Financial analysis, 2008–2013

Ratios	2008	2009	2010	2011	2012	2013
Liquidity (general)	0.56	0.56	0.69	0.61	0.50	0.41
Current ratio	1.02	1.32	1.88	1.78	1.70	1.49
Quick ratio	0.50	0.74	1.30	1.09	1.03	0.90
Working capital	1.018	18.645	49.851	52.952	48.482	40.826
Working capital—needed	−11.808	−10.598	−10.084	−7.956	−11.981	−11.001
Treasury	12.826	29.243	59.935	60.908	60.463	51.827
Short term debt	0.21	0.17	0.11	0.11	0.10	0.11
Long term debt	0.30	0.37	0.29	0.33	0.39	0.43
Total debt ratio	0.52	0.53	0.40	0.45	0.49	0.54
Financial debit ratio	0.22	0.29	0.23	0.26	0.29	0.36
D/E	0.45	0.62	0.38	0.47	0.57	0.77
Gross margin	0.34	0.40	0.36	0.32	0.25	0.23
Operational margin	0.21	0.25	0.22	0.19	0.12	0.11
Net profit margin	0.15	0.16	0.17	0.14	0.07	0.08
ROA	0.16	0.13	0.09	0.08	0.05	0.05
ROE	0.23	0.18	0.12	0.10	0.06	0.07
Days sales in inventory	51	71	52	61	51	51
Days sales outstanding	25	28	29	33	29	27
Operational cycle	76	98	82	94	80	78
Financial cycle		98	82	94	80	78

Source: Adaptation of the Petrobras (2009, 2010, 2011, 2012, 2013, 2014)

Fig. 1 Distribution of stock prices, 2000 till 2014. Source: Adaptation of the Petrobras (2009, 2010, 2011, 2012, 2013, 2014)

Table 2 Financial analysis, 2014–2015

Petrobras ratios:	2014	2015/1
Liquidity (general)	0.38	0.37
Current Ratio	1.63	1.52
Quick Ratio	1.09	0.99
Short term debt	0.10	0.11
Long term debt	0.50	0.52
Total debt	0.61	0.63
Financial debt	0.44	0.48
D/E	1.13	1.31
Gross margin	0.24	0.30
Operating margin	−0.06	0.18
Net profit margin	−0.06	0.07
ROA	−0.03	0.02
ROE	−0.07	0.02
Days sales in inventory	23	100
Days—operational cycle	65	322
Days—financial cycle	65	322

Source: Adaptation of Petrobras (2014, 2015)

company (Petrobras 2015) on December 31, 2014. However, this amount is included in the financial statements as "other operating income". This account of "other operating income", has a total of US$75 billion in expenses and cannot be identified in the notes of the financial report of company (Petrobras 2015). After nearly a year of review, Petrobras announced the balance of 2014 and also the first quarter of 2015. It was then performed an analysis of the financial statements according to Table 2.

The notes do not mention in which assets exactly these values were applied. If verified fixed assets, investments and intangible assets, do not identify the loss of numbers. The account under which there was a great increase, and can be identified by analysing the indexes, is the indebtedness of the Company, in relation D/E. That is because the company's assets decreased. Thus, if an analyst checks the company's financial statements, despite the loss of 44 billion in impairment, there were no major changes in its financial statements.

5 Discussion

While all this news of movement occurred between the Brazilian presidential election, the disclosure of corruption installed in Petrobras and disclosure of corrected statements, the market actively reacted in relation to non-disclosure of financial information. Figure 2 shows the evolution of the share price between September 2014 and June 2015.

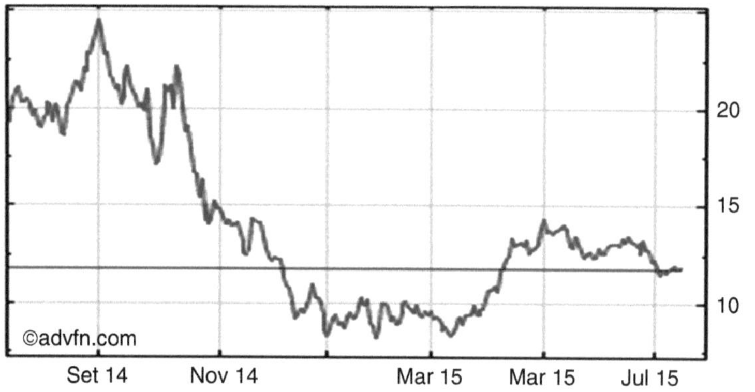

Fig. 2 Distribution of stock prices, 2014 till 2015. Source: Adaptation of the Petrobras (2014, 2015)

One cannot compare Petrobras over time. If there is no trust in the information, and an operating expense of US$ 75 billion was launched in 2014, how can you compare this figure with previous years? How do you earn your income? Thus, with this reaction of investors it is important to note the qualitative characteristics of accounting information.

The relevance of information: a loss of US$ 44 billion impairment is significant enough that there explanatory notes to identify exactly what happened to the assets to which this value was pegged. At the same time, use the figure of US$ 75 billion as other operating expenses does not seem to be the best alternative in accounting disclosure.

According to Petrobras (2015), a faithful representation: the items of the financial statements should show the nearest value of reality and their recoverability and:

- Timeliness: the recoverability of loss of information assets is late. What is the use of this information only to its publication at this time?
- Verifiability: as the investor can verify this information? The audit had access to all these assumptions? There was enough time to make all necessary checks?
- Comparability: how these statements may, from now on, be comparable? The company should have made an adjustment for prior years in all financial statements for investors to know exactly at what time the loss was identified.
- Understandability: it is not possible for ordinary investors to identify what really happened in the assets of Petrobras. Take, for example the table showing the movement of the fixed assets of the company.

Since 2010, they were implemented international accounting standards. There is evidence that the value of the loss impairment was found only at this time. Why the company did not publish these losses over the past 5 years?

6 Conclusion

The relevance of the research is due to the most important news that appears about the Petrobras. Indeed, the originality of the paper is higher, because the unethical behavior is always very hard to prove and especially at this level.

The main contribution of this research is to explore the accounting communication with the society made by Petrobras as the largest company in Latin America and one of the biggest integrated energy companies in the world, with a strategic role in oil geopolitics.

The authors aim to understand the relevance of Petrobras activity through the accounting processes and the financial report that it is strongly regulated as a fundamental area of the financial markets. This company had gained many prizes as one of the best accounting practices in the country. In addition, however, we may understand that the evidences shows otherwise.

The discussion that the authors would like to begin, in this case, is how much could the government, the directors and the auditors influence in the accounting practices? Already, Hendriksen and Van Breda (1999) explore this discussion and the authors understand that this is not an easy task for the companies.

The main limitations are the application of just one case study, in on company that is calling for attention in Brazil because of the corruption inside the board. A case study and descriptive analysis were done in order to understand what the accounting information brings to the investors and also the availability of the information is still not easy for collect and this analysis is court decision, which increase the responsibility of the discussion made in the paper.

As the future development, the authors aim to widespread the accounting and financial report as weapon to combat mal-functioning of the financial markets and help several stakeholders.

Acknowledgements The authors wish to thank Prof. José Ángel Pérez López of Universidad de Seville (Spain) and the assistance of the anonymous reviewers whose insightful comments and suggestions enabled us to improve this research. The Portuguese authors are full member of the Centro de Investigação de Contabilidade e Fiscalidade (CICF-IPCA) and Brazilian are full member of Mackenzie Research. Ideas expressed in the article are those of the authors and should not be attributed to any organization.

References

Ahmed, K., & Nicholls, D. (1994). The impact of non-financial company characteristics on mandatory disclosure compliance in developing countries: The case of Bangladesh. *International Journal of Accounting, 29*(1), 62–77.

Antunes, J., Antunes, G. M. B., & Penteado, I. M. (2007). *A convergência contábil brasileira e a adoção das normas internacionais de contabilidade: o IFRS-1.* São Paulo: X SEMEAD—Seminários em Administração FEA-USP.

Arens, A., Elder, R. J., & Beasley, M. (2012). *Auditing and assurance services: An integrated approach*. London: Pearson Prentice Hall.
Ball, R., & Shivakumar, L. (2006). The role of accruals in asymmetrically timely gain and loss recognition. *Journal of Accounting Research, 44*(2), 207–242.
Baptista, E. M. B. (2009). Ganhos em transparência versus novos instrumentos de manipulação: o paradoxo das modificações trazidas pela Lei 11.638. *Revista de Administração e Empresas, 49* (2), 234–239.
Bushman, R., Chen, Q., Engel, E., & Smith, A. (2004). Financial accounting information, organizational complexity and corporate governance systems. *Journal of Accounting and Economics, 37*(2), 167–201.
Buzby, S. L. (1974). Selected items of information and their disclosure in annual reports. *The Accounting Review, 49*(3), 423–436.
Chen, K. C. W., Chen, Z., & Wei, K. C. J. (2004). *Disclosure, corporate governance and the cost of equity capital in emerging markets*. Working paper series. Available: http://ssrn.com/abstract=422000
Conselho Federal de Contabilidade (CFC). (2005). *Comitê de Pronunciamentos Contábeis*. São Paulo: CFC.
Cooke, T. (1989). Disclosure in the corporate annual reports of Swedish companies. *Accounting and Business Research, 19*(74), 113–124.
Costa, T. A., Almeida, S. R. V., & Silva, A. H. C. (2013). Uma análise dos impactos da adoção das normas internacionais de contabilidade no Brasil: um estudo de empresas de capital aberto no setor de bebidas, alimentos e comércio. *Pensar Contábil, 15*(56), 3.
DeAngelo, L. E. (1981). Auditor independence, 'low balling' and disclosure regulation. *Journal of Accounting and Economics, 3*(2), 113–127.
Dechow, P. M., & Schrand, C. M. (2004). *Earnings quality*. Charlottesville: CFA Institute.
Domenico, D. D., Dal Magro, C. B., & Klann, R. C. (2014). Impactos da adoção das normas contábeis internacionais completas (full ifrs) nos indicadores econômico-financeiros de empresas listadas na Bovespa. *Revista de Informação Contábil, 8*(1), 52–70.
Government of Brazil, Ministry of Finance. (2007). *Lei 11.638, de 28 de dezembro de 2007. Altera e revoga dispositivos da Lei n. 6.404, de 15 de dezembro de 1976, da Lei n. 6.385, de 7 de dezembro de 1976 e estende às sociedades de grande porte disposições relativas à elaboração e divulgação de demonstrações financeiras*. Disponível em: http://www.planalto.gov.br/ccivil_03/_ato2007-2010/2007/lei/11638.htm
Hendriksen, E., & Van Breda, M. (1999). *Teoria da Contabilidade*. São Paulo: Atlas (Tradução da 5a edição americana por Antonio Z. Sanvicente).
Herculano, H. A., & Moura, G. D. (2015). Informação contábil e concentração acionária: análise sob a ótica da persistência e da oportunidade. *Revista Ambiente Contábil, 7*(2), 231.
Hossain, M. (2000). *The disclosure of information in the annual reports of financial companies in developing countries. The case of Bangladesh*. Master's Thesis of Accounting and Finance. Manchester: The University of Manchester.
Hossain, M., Tan, L. M., & Adams, M. B. (1994). Voluntary disclosure in an emerging capital market: Some empirical evidence from companies listed on the Kuala Lumpur stock exchange. *International Journal of Accounting, 29*(4), 334–351.
Inchausti, B. (1997). The influence of company characteristics and accounting regulations on information disclosure by Spanish firms. *European Accounting Review, 6*(1), 45–68.
International Accounting Standards (IASB). (2015). *International accounting standards*. New York: IASB.
International Federation of Accountants (IFAC). (2015). *Manual das Normas Internacionais de Controlo de Qualidade, Auditoria, Revisão, Outros Trabalhos de Garantia de Fiabilidade e Serviços Relacionados*. Lisboa: OROC.
Kahl, A., & Belkaoui, A. (1981). Bank annual report disclosure adequacy internationally. *Accounting and Business Research, 43*(11), 189–196.

Maia, H. A. (2011). *Impacto das características específicas das companhias abertas brasileiras no nível de evidenciação contábil durante o processo de convergência às normas internacionais de contabilidade. Dissertação do Mestrado Profissional em Controladoria Empresarial*. São Paulo: Universidade Prestiberiana Mackenzie.

Maia, H. A., Formigoni, H., & Silva, A. A. (2012). Empresas de auditoria e o compliance com o nível de evidenciação obrigatório durante o processo de convergência às normas internacionais de contabilidade no Brasil. *Revista Brasileira de Gestão de Negócios, 14*(44), 335–352.

Malone, D., Fries, C., & Jones, T. (1993). An empirical investigation of the extent of corporate financial disclosure in the oil and gas industry. *Journal of Accounting, Auditing and Finance, 8* (3), 251–273.

Martins, V. G., Oliveira, A. S., Niyama, J. K., & Diniz, J. A. (2014). Níveis diferenciados de governança corporativa e a qualidade da informação contábil durante o processo de convergência às normas internacionais de contabilidade. *Revista ConTexto, 14*(27), 23–42.

Martins, O. S., & Paulo, E. (2010). Reflexo da adoção das IFRS na análise de desempenho das companhias de capital aberto no Brasil. *Revista de Contabilidade e Organizações, 4*(9), 30–54.

Messier, F. W., Glover, S. M., & Prawitt, F. D. (2008). *Auditing and assurance services a systematic approach*. New York: McGraw Hill.

Mohamed, Z. M., & Janggu, T. (2006). Corporate social disclosure of construction companies in. *Malaysia, 5*(1), 85–114.

Moraes, M. V. M., Macedo, M. A. S., & Sauerbronn, F. F. (2014). *O impacto da adoção das IFRS no resultado contábil das pessoas jurídicas no Brasil*. São Paulo: XIV Congresso USP de Controladoria e Contabilidade.

Moura, G. D., Franz, L., & Cunha, P. R. (2013). *Qualidade da informação contábil em empresas familiares: influência dos níveis diferenciados de governança da BM&FBOVESPA, tamanho e independência do conselho de administração*. Anais SIMPOI.

Murcia, F. D.-R., Rover, S., Lima, I., Fávero, L. P., & Lima, G. (2008). "Disclosure verde" nas demonstrações contábeis: características da informação ambiental e possíveis explicações para a divulgação voluntária. *Revista Universo Contábil, 11*(1–2), 260–278.

Ng, E. J., & Koh, H. C. (1993). Companies with non-mandatory accounting pronouncements: The Singapore experience. *Singapore Management Review, 15*(1), 41–55.

Ohlson. (1995). Earnings, book values, and dividends in equity valuation. *Contemporary Accounting Research, 11*(2), 661–687.

Paulo, E., & Martins, E. (2007). *Análise da qualidade das informações contábeis nas companhias abertas*. Rio de Janeiro: XXXI Encontro da ANPAD.

Peng, S., Tondkar, R. H., Smith, J. V. D. L., & Harless, D. W. (2008). Does convergence of accounting standards lead to the convergence of accounting practices? A study from China. *International Journal of Accounting, 43*(4), 448–468.

Petrobras. (2009). *Relatório e Contas 2008*. Cidade: Petrobras.

Petrobras. (2010). *Relatório e Contas 2009*. Cidade: Petrobras.

Petrobras. (2011). *Relatório e Contas 2010*. Cidade: Petrobras.

Petrobras. (2012). *Relatório e Contas 2011*. Cidade: Petrobras.

Petrobras. (2013). *Relatório e Contas 2012*. Cidade: Petrobras.

Petrobras. (2014). *Relatório e Contas 2013*. Cidade: Petrobras.

Petrobras. (2015). *Relatório e Contas 2014*. Cidade: Petrobras.

Ponte, V. M. R., Freitas, J. B., & Oliveira, M. C. (2013). *Impactos da adoção das IFRS nas demonstrações consolidadas dos bancos listados na BM&FBovespa*. Rio de Janeiro: XXXVII Encontro da Anpad.

Santiago, J. S., & Cavalcante, P. R. N. (2014). *A qualidade da informação contábil e o processo de convergência internacional nas empresas de capital aberto do setor de construção no Brasil*. Rio de Janeiro: VIII Congresso AnpCONT.

Silva, R. L. M. (2013). *Adoção completa das IFRS no Brasil: qualidade das demonstrações contábeis e o custo de capital próprio. Tese da Faculdade de Economia, Administração e Contabilidade*. São Paulo: Universidade de São Paulo.

Singhvi, S. S., & Desai, H. B. (1971). An empirical analysis of the quality of corporate financial disclosure. *Accounting Review, 46*(1), 129–138.

Tay, J. S. W., & Parker, R. H. (1990). Measuring international harmonization and standardization. *Abacus, 26*(1), 71–88.

Uwuigbe, U. (2011). An empirical investigation of the association between firms' characteristics and corporate social disclosures in the Nigerian financial sector. *Journal of Sustainable Development in Africa, 13*(1), 60–74.

Vieira, R. B. (2010). *Impactos da implantação parcial das IFRS no Brasil: efeitos na qualidade das informações contábeis das empresas de capital aberto. Dissertação da Faculdade de Economia, Administração e Contabilidade*. São Paulo: Universidade de São Paulo.

Wallace, R. S. O. (1987). *Disclosure of accounting information in developing countries: A case study of Nigeria*. Ph.D. Thesis Economics Department. Exeter: University of Exeter.

Wallace, R. S. O., & Naser, K. (1995). Firm-specific determinants of the comprehensiveness of mandatory disclosure in the corporate annual reports of firms listed on the stock exchange of Hong Kong. *Journal of Accounting and Public Policy, New York, 14*(4), 311–368.

Warren, C. S., Reeve, J. M., & Fess, P. E. (2014). *Contabilidade gerencial*. São Paulo: Pioneira Thomson Learning.

Yin, R. (2012). *Case study research: Design and methods*. Thousand Oaks: Sage Publications.

Mineral Supply Chain Transparency: Soft and Hard Laws on Supply Chains Due Diligence and the Rise of Public-Private Partnerships

Fabiana Di Lorenzo

1 Introduction: The International Debate on Human Rights and Business and Mandatory and Voluntary Legal Instruments on Due Diligence and Responsible Sourcing of Minerals

The subject of human rights has historically being linked to States, as these have always been considered the primary abusers of human rights (Green and Ward 2004). However, following a number of corporate 'misconduct' episodes—Bhopal (1984), the explosion of Ford Pinto cars due to the faulty design of the fuel system (1970s), legal action against Shell for the detention and then execution of nine Ogoni activists in the Niger Delta (1995), the Rana Plaza collapse (2013)—civil society and academics eventually turned their attention towards the interplay between human rights and business, and started looking at the negative impact of business on human rights (Bernaz 2017; Baumann-Pauly and Nolan 2016; Deva 2012; Kaeb 2008). The focus on 'negative' impact was "not [brought forth] because business is evil; rather, it is because human rights, as a field of study, tends to focus on violation" (Bernaz 2017, p. 8).

The international debate on human rights and business kicked off in 2007, when a series of consultations on the subject were convened on behalf of Professor John Ruggie, Special Representative of the Secretary-General of the United Nations. One of the series of consultations called for multi-stakeholder or public private partnership (PPP) initiatives to improve business' human rights performance. The consultation concluded that, whilst States have the duty to protect against human rights violations, business can also play a significant role in addressing the risks of human rights abuses linked to business activities

F. Di Lorenzo (✉)
Estelle Levin Limited, Cambridge, UK
e-mail: fabiana@estellelevin.com

H. Lu et al. (eds.), *Building New Bridges Between Business and Society*, CSR, Sustainability, Ethics & Governance, DOI 10.1007/978-3-319-63561-3_9

(Secretary General of the United Nations Consultation 2007). These principles would later be formally adopted as the UN Guiding Principles on Business and Human Rights (UN Guiding Principles) drafted by UN Special Representative on business and human rights John Ruggie, and formally endorsed by the Human Rights Council in 2011. The UN Guiding Principles outlined the process by which companies fulfil their responsibility to respect human rights (UN Guiding Principle No. 15). The process involves the adoption of an internal human rights policy, and conducting due diligence to identify, prevent, mitigate and account for how they address their impact on human rights. During this same period, the Organisation for Economic Co-operation and Development (OECD) reviewed its Guidelines on Multinational Enterprises (MNsEs) and aligned them with the principles of due diligence outlined in the Guiding Principles. The OECD then proceeded to create dedicated guidelines for different types of businesses in agriculture, extractive sector, financial sector, mineral supply chains, and textile and garment supply chains. This article will solely refer to the OECD Guidelines on mineral supply chains.

The UN Guiding Principles on Business and Human Rights are organised around three pillars: 1. State duty to *protect* against human rights abuses, 2. corporate responsibility to *respect* human rights, and 3. access to *remedy* by victims of corporations. The UN Guiding Principles provided a common basis and a minimum set of standards and processes to businesses willing to improve their human rights performance. All businesses can refer to them. The State has a duty to protect against human rights abuses, and it is emphasised in the Principles that this duty also includes creating an operative environment where businesses can respect human rights. This can only take place when legal efficiency and enforcement mechanisms are in place (Cotterrell 1984; Di Lorenzo 2014). Efficiency indicates the provision and oversight of public service, such as building roads, keeping people safe, and providing education and health services. Enforcement, on the other hand, refers to establishing measures and supervision mechanisms to ensure compliance with the law (Cotterrell 1984; Di Lorenzo 2014).

In the late 1990s the major campaign carried out by civil society against so-called blood diamonds, which were fueling conflicts in Angola (Human Rights Watch 1994), marked the beginning of public interest towards the links between conflicts and the exploitation of minerals in high risk and conflict affected countries. In 2010 a set of laws was adopted to promote the responsible sourcing of minerals (tin, tantalum, tungsten and gold, also referred as the 3TG) from high risk and conflict affected areas. Legislative and regulatory initiatives seek to break the links between minerals and armed groups, and promote conflict mineral-free supply chains through effective risk assessment, mitigation, remediation, 'know your customer' and 'know your product' measures and reporting mechanisms. All aim at promoting good governance and due diligence to ensure that the proceeds from the extraction of and trade in minerals do not contribute to aggravating and perpetuating the violation of human and environmental rights. In conflict-affected and high risk areas the proceeds from the extraction of and trade in minerals can aggravate and perpetuate the violation of human rights in conflict zones. When sourcing minerals, companies, either deliberately or unintentionally, run into the risk of directly or indirectly supporting armed groups, which illegally take control over mines and their trading routes, and use the proceeds to finance their activities. They also run the risk of unwillingly be involved in cases of bribery, corruption, fraud, forced labour, or human rights abuses in the mineral supply chain. For this reason, new regulatory frameworks have been introduced that require companies to report on the sources of minerals and carry out due diligence on the upstream (from the mines to the smelters and refiners) or the downstream section of supply chains (from the smelters and refiners to the manufacturers).

2 Oscillating Between Mandatory and Voluntary Legal Frameworks on Mineral Supply Chain Due Diligence

As aforementioned, on 16 June 2011, the Human Rights Council endorsed the Guiding Principles on Business and Human Rights: Implementing the United Nations "Protect, Respect and Remedy" Framework (UN Human Rights Council A/HRC/RES/17/4). The UN Guiding Principles reinforce the idea that the State is the "bedrock" of human rights protection (Ruggie 2013). Corporations, on the other hand, have the duty to respect human rights by taking measures to identify, prevent, mitigate and account for how they manage negative impacts on human rights (Guiding Principle No. 17). The UN Guiding Principles are a voluntary form of law. However, as Ruggie (2013) and Bernaz (2017) pointed out, they still brought the notions of corporate responsibility and law closer together. When in 2010 the US government decided to regulate the activities of companies sourcing 3TG minerals through the Dodd-Frank Wall Street Reform and Consumer Protection Act (the Dodd-Frank Act) (Dodd-Frank 2010), section 1502 on Conflict Minerals, the goal was to regulate minerals originating from conflict-affected countries. The law requires US-listed companies to disclose publicly their use of "conflict minerals" originating from the Democratic Republic of Congo (DRC) or surrounding areas (the Great Lake Region), to file a Conflict Minerals Report, and to exercise due diligence on the source and chain of custody of their conflict minerals. If the company knows or has reason to believe that the minerals may have originated from the DRC or surrounding countries (the Covered Countries), and knows or has reason to believe that the minerals may not be from scrap or recycled sources, then the company must undertake due diligence as to the source and chain of custody of the minerals, file a Conflict Minerals Report and publish it on the company's website. "The Conflict Minerals Report must include the following information: (1) *Due Diligence*: A description of the measures the registrant has taken to exercise due diligence on the source and chain of custody of those conflict minerals; (i) The registrant's due diligence must conform to a nationally or internationally recognized due diligence framework, if such a framework is available for the conflict mineral" (SEC guidance). In 2012, the US Securities and

Exchange Commission (SEC) recognised the OECD Due Diligence Guidance for Responsible Supply Chains of Minerals from Conflict-Affected and High Risk Areas (OECD Guidance) as an international framework for companies that were required to comply with the Dodd-Frank legislation. The OECD Guidance has hence become the framework of reference of any voluntary and mandatory law on supply chain transparency and due diligence on mineral supply chains. It provides five steps for risk-based due diligence on mineral supply chains and it applies to all raw materials and to all regions. It is beyond the scope of this article to review in detail each step, however a short overview is here provided. The first step relates to establishing a strong management system; the second step, undertaking risk assessment, which can be carried out individually or in cooperation with customers; the third step refers to the need to mitigate risks that have been identified and regularly monitor them; the forth step, to taking part in audit programmes; and the last step defines reporting the efforts made in due diligence on an annual basis. At the time when this article is being written, US President Trump has expressed his intention to cancel Section 1502 of the Dodd-Frank Act, and questions are being raised over the benefits of the Rule overall (IRIN News 2017). As a result, the SEC is running a consultation to identify a better approach for the responsible sourcing of 3TG (State Department 2017). Chances that Dodd-Frank Act's Section 1502 would remain as it is appear low.

The 2010 Dodd-Frank Act led the way to further legislation on global supply chain transparency in relation to human and labour rights to prevent the entry of conflict minerals into the American and later on into the European market. Nonetheless improvements to the Act as it stands today could be made in a few areas. This includes the geographic scope of the Act: its regional coverage is restrictive to the DRC, which is restricted to the DRC and surrounding countries, thereby neglecting other high risk countries such as Afghanistan, Colombia and Myanmar, where the trade of precious minerals and stones contributes to the perpetuation of conflicts (Global Witness 2014). Furthermore, issues have been raised in relation to its enforcement and the challenges faced by companies to fully comply with it. Despite being a form of mandatory regulation, 80% of public companies in the US have failed to meet the minimum requirements of the US conflict minerals law and to map their supply chain and minerals purchased by them (Global Witness 2015; Sarfaty 2015). The Section 1502 of the Dodd-Frank Act is currently undergoing a consultation to determine its future and some industry leaders hope the new rule will incorporate lessons that have been learned over the last years, instead of being simply 'scrapped'.[1]

While the USA is facing uncertainty over the future of its conflict mineral legislation, the European Parliament adopted the Regulation (EU) 2017/821 of the European Parliament and of the Council laying down supply chain due diligence obligations for Union importers of tin, tantalum and tungsten, their ores, and gold

[1]Opinion gathered through informal conversations with two sustainability leaders working in major downstream companies.

originating from Conflict-Affected and High-Risk Areas on 16 March 2017 commonly referred to as the Conflict Minerals Regulation. The two mandatory laws have some similarities but also important differences. The EU conflict minerals rule—just as the one in the USA—aims at breaking the link between minerals and armed groups (European Council press release 2017) and they are both aligned with the OECD five steps of due diligence. However, whilst Dodd-Frank rule targeted primarily downstream companies, the EU regulation targets importers, refiners and smelters, which constitute the pinch point of the supply chain to ensure transparency for downstream companies. In line with the new regulation, EU importers, refiners and smelters will have to carry out due diligence as of 1 January 2021. Like Dodd-Frank Section 1502, the EU regulation is focusing on the 3TG. However it has a global geographical scope: it requests due diligence on 3TG to be carried out on all minerals coming from any conflict-affected and high risk areas. While the focus on a larger geographical scope was welcomed by civil society, some saw a missed opportunity as the Regulation failed to extend to other types of minerals (for instance jade, cobalt and other minerals exported from high risk and conflict-affected areas) (Global Witness 2014; London Mining Network 2015). The OCED Guidance itself recognised its framework of due diligence to be applicable to all minerals. Interestingly, the EU Regulation is potentially aligned with the OECD Guidance on this point and it does contain a provision (art. 24), which could allow the Commission and the High Representative of the Union for Foreign Affairs and Security Policy to propose further mandatory measures.

Inspired by voluntary regulation, the Chamber of Commerce of Metals, Minerals and Chemicals Imports and Exports (CCCMC), supervised by the Ministry of Commerce, introduced in October 2014 the Chinese Due Diligence Guidelines for Responsible Mineral Supply Chains, which align due diligence carried out by Chinese companies with international standards. This voluntary provision covers a broad range of issues such as labour and human rights, environmental protection, supply chain due diligence, corruption and community engagement. Yet for the time being it remains a means of soft law with limited space for full enforcement and still requires intense effort from the CCCMC to promote its adoption by Chinese companies.

In this journey which oscillates between voluntary and mandatory legislation on global mineral supply chains, the industry associations have played a crucial role in promoting responsible sourcing of minerals. They actively supported their members by introducing reporting tools, upstream traceability and due diligence solutions and training. Industry associations such as the Electronic Industry Citizenship Coalition, The Global Tin Industry Association and the Tantalum-Niobium International Study Center, have helped companies join forces to find synergies and reduce to the minimum the risks of duplications. Similarly, it is again industry associations who are leading the way in setting new standards which go beyond existing mandatory regulations (see for example the creation of the Advance Responsible Sourcing of Raw Materials in Technology Supply Chains). By going beyond the 3TG, industry initiatives are playing a key leadership role in the international debate on global supply chain transparency by seeking "to address a broader suite of social and environmental impacts along global supply chains in raw materials" (Di Lorenzo 2017). Along with the work of industry associations, a number of PPPs—in the form of tripartite social cross-sector partnerships (Stadtler 2012)—have been established all over the world to promote responsible sourcing practices. Their purpose is to find durable solutions for the promotion of responsible and transparent mineral supply chains. These PPPs bring together private, public and civil society organisations to address complex issues which each cannot solve on their own. The momentum created by these partnerships is changing the way human rights are pushed forward, and places PPPs in a key role for the advancement of human rights in global supply chains.

3 Mineral Supply Chains Due Diligence and the Rise of Public Private Partnerships for Legal Advancement of Human Rights

Voluntary and mandatory legislations to regulate the responsible sourcing of minerals have been accompanied by a rise in the number of so-called PPPs and multi-stakeholder collaborations. With the rise of PPPs, private companies, civil society and public actors join forces to promote compliance with human rights, environmental standards and laws (Di Lorenzo 2014). They also provide the platform for different actors to share knowledge, overcome the due diligence gap created by poor domestic governance in producing countries (Sarfaty 2015) and to identify solutions to the complex challenge of responsible sourcing of minerals. Public and private actors do resort to these new forms of institutional organisations when their supply chains and operations reach out to countries with weak governance, high level of corruption and poor law enforcement. In these instances, PPPs are seen as the best solution to protect human and labour rights (Bäckstrand 2006; Baumann-Pauly and Nolan 2016; Di Lorenzo 2014; Scherer and Palazzo 2011). This is because public, private and not-for-profit organisations work together to share knowledge, promote responsible sourcing solutions, and to improve standards of living, create better lives, and respect *and* protect human rights. Through the PPPs model, companies become political actors, as well as agents of human rights advancement, legal enforcement and legal efficiency (Di Lorenzo 2014). This transition is reinforced by the UN Guiding Principles, which call on companies to "respect" human rights but also in joining multi-stakeholder initiatives (Jochnick 1999; Ruggie 2008) to tackle issues that cannot be overcome if operating on their own.

As aforementioned, the momentum created by the PPPs is essentially changing the way human rights are advanced for. Human rights *protection,* which used to be the sole duty of States, is now also being required from the private sector (Crane and Matten 2007; Crane et al. 2008; Di Lorenzo 2014). This requirement is not merely to ensure companies *respect* human rights and perform due diligence, but also that they participate in public policy and legal advancement through PPPs' initiatives (Scherer and Palazzo 2011). In the mineral sector, PPPs members work together to identify common strategies for due diligence, promote compliance, improve governance in upstream countries and support the livelihoods of miners. Examples of these platforms are the Public-Private Alliance for Responsible Minerals Trade (PPA), the IDH Tin Working Group, the European Partnership for Responsible Minerals (EPRM), the Responsible Cobalt Initiative, the Responsible Mica Initiative, and the Responsible Battery Alliance. The goals of these PPPs are diverse, but all aim at promoting responsible mining and sourcing practices, including improving the life standards of artisanal, small-scale miners and governance in producing countries. However, the success of these PPPs is threatened by the weak capacity of government in sourcing countries to properly develop their mineral sectors and set the conditions for due diligence to be implemented in a successful manner. For instance, an evaluation of the PPA revealed that even when governments such as the DRC commits to implementing due diligence, at the implementation stage government officials still demonstrate a lack of technical understanding of what due diligence is and how to go about its implementation (OECD 2013). Another issue in relation to proper implementation of due diligence in sourcing countries with low governance relates to the government's inability to establish a strategic communication plan whose messages reach out to citizens and business alike in a way that educates them on what is expected from them. For instance, in the Great Lake Region, the exporters are subject to audits. These audits examine the mineral supply chains from the point of export all the way to the point of origin of the minerals, including all actors involved in mining, buying, selling, transporting and handling of the minerals. Mine sites are inspected annually by a government mines inspector and such inspections are then validated by annual Independent Third Party Audit. Informal conversations with businesses and experts on the ground revealed how some companies subject to audits still lack the capacity and proper information on what it is expected from them to fully comply with due diligence requirements.

As PPPs find broad acceptance in the literature on corporate social responsibility and sustainable development, these new forms of institutional organisation are gaining recognition as being the most appropriate means for human rights protection and advancement along global supply chains (Ashman 2001; Utting 2002; Zammit 2003; Murphy and Bendell 1999). Traditionally, PPPs have worked on the development and implementation of compliance with human and environmental rights and standards (see for instance the Fair Labor Association) (Baumann-Pauly and Nolan 2016). But PPPs have also gone beyond the focus on compliance when different parties decided to undertake 'specific tasks, share risks, responsibilities, resources, competencies and benefits' (Ruggie 2008; Bäckstrand 2006), in the belief that States have failed in complying with their obligations and only a coordinated strategy can effectively advance human rights (Bäckstrand 2006; Scherer and Palazzo 2011), promote livelihoods and protect companies' reputation (Pedersen and Andersen 2006; OECD 2015; Roberts 2003; Utting 2005). Well-known PPPs such as those in cocoa, cotton and tobacco supply chains, which have been operational for far more than a decade, have demonstrated the risks of failure that come with neglecting the importance of holding governments in producing countries[2] accountable over the delivery of vital services and support to their citizens—the ultimate guarantee for real human rights advancement through legal enforcement and also efficiency (Di Lorenzo 2013). This is a lesson that PPPs in the mineral sector should learn from and avoid replicating. To conclude, to be effective PPPs require States in sourcing countries to play a considerably greater role in legal enforcement and efficiency. This can be better achieved by inviting representatives of those governments to sit on PPPs' boards—and not only *engaging* with them. This would create the opportunity and (moral) obligation for different stakeholders to hold them accountable for the work carried out with the taxes collected from miners, as well as money received from PPPs' initiatives. Companies alone cannot succeed in doing due diligence and mapping their supply chain without this aspect being given proper consideration.

4 Conclusions

This article has outlined the diverse legal instruments mandated to increase transparency on mineral supply chains and their global operations through reporting requirements. It delved into the specificities of each legal instrument and their scope. The article then highlights the position of industry associations and how these have played a role in supporting and promoting their members' compliance with due diligence laws. The final section of this article is dedicated to the rise of PPPs through which companies, governments and civil society organisations are working together to improve supply chain management and promote due diligence, while also investing in alternative livelihood and community development programmes for miners. PPPs are now perceived by all parties to be the most appropriate means to advance legal compliance and protect human rights. At a time when multi-stakeholder platforms are being called to fill in the vacuum left by government in dispensing public services (Scherer and Palazzo 2011), PPPs should focus their efforts in holding governments of producing countries accountable for legal enforcement and services offered to mining communities. This goes beyond regulation and requires governments to perform a strong role, as dispensers of public goods and sustainable providers of public services for local communities, fundamental to fostering responsible sourcing, and the promotion and respect of human rights.

[2]Personal communication with a manager in a major international organisation and a manager in an international trade association.

References

Ashman, D. (2001). Civil society collaboration with business: Bringing empowerment back. *World Development, 29*(7), 1097–1113.

Baumann-Pauly, D., & Nolan, J. (2016). *Business and human rights from principles to practice*. Oxon: Routledge.

Bäckstrand, K. (2006). Multi-stakeholder partnerships for sustainable development: Rethinking legitimacy, accountability and efficiency. *European Environment, 16*, 290–306.

Bernaz, N. (2017). *Business and human rights. History, law and policy – Bridging the accountability gap*. Oxon: Routledge.

Chinese Due Diligence Guidelines for Responsible Mineral Supply Chains. http://mneguidelines.oecd.org/chinese-due-diligence-guidelines-forresponsible-mineral-supply-chains.htm

Cotterrell, R. (1984). *The sociology of law: An introduction*. London: Butterworth.

Crane, A., & Matten, D. (2007). *Business ethics. Managing corporate citizenship and sustainability in the age of globalisation*. Oxford: Oxford University Press.

Crane, A., Matten, D., & Spence, J. (2008). *Corporate social responsibility. Readings and cases in global context*. New York: Routledge.

Deva, S. (2012). *Regulating corporate human rights violations—humanizing business*. New York: Routledge.

Di Lorenzo, F. (2013). The role of public private partnerships in labour rights advancement. *The Oxford Human Rights Hub*. Retrieved April 1, 2015, from http://ohrh.law.ox.ac.uk/the-role-of-public-private-partnerships-in-labour-rights-advancement/

Di Lorenzo, F. (2014). *All is quiet on the partnerships front: Scrutinising power dynamics in cocoa partnerships and their effects on child rights advancement in Ghana*. PhD thesis. London: King's College. http://ethos.bl.uk/OrderDetails.do?uin=uk.bl.ethos.650763

Di Lorenzo, F. (2017). *Dodd-Frank Section 1502: The future of due diligence if Dodd-Frank is repealed under the new presidency*. Retrieved May 19, 2017, from http://www.estellelevin.com/dodd-frank-1502-future-due-diligence/

Dodd-Frank Wall Street Reform and Consumer Protection Act. (2010). Retrieved April 2, 2015, from http://www.cftc.gov/ucm/groups/public/@swaps/documents/file/hr4173_enrolledbill.pdf

European Council Press Release. (2017). *Conflict minerals: Council adopts new rules to reduce financing of armed groups*. Retrieved April 26, 2016, from http://www.consilium.europa.eu/en/press/press-releases/2017/04/03-conflict-minerals/

Global Witness. (2014). *The case for human rights due diligence*. Retrieved May 30, 2015, from https://www.globalwitness.org/documents/13425/psddbrief.pdf

Global Witness. (2015). *Digging for transparency. How US companies are only scratching the surface of conflict mineral reporting*. Retrieved June 20, 2015, from https://www.globalwitness.org/campaigns/democratic-republic-congo/digging-transparency/

Green, P., & Ward, T. (2004). *State crime*. London: Pluto Press.

Human Rights Watch. (1994). *Angola: Arms trade and violations of the Laws of War since the 1992 elections*. New York: Human Rights Watch.

IRIN News. (2017). *Who pays the hidden price for Congo's conflict-free minerals?* Retrieved April 26, 2017, from http://www.irinnews.org/investigations/2017/02/14/who-pays-hidden-price-congo%E2%80%99s-conflict-free-minerals?utm_source=IRIN
+-+the+inside+story+on+emergencies&utm_campaign=ab59f9f88d-RSS_
EMAIL_ENGLISH_ALL&utm_medium=email&utm_term=0_
d842d98289-ab59f9f88d-15761377

Jochnick, C. (1999). Confronting the impunity of non-state actors: New fields for the promotion of human rights. *Human Rights Quarterly, 21*(1), 56–79. http://heinonline.org/HOL/LandingPage?handle=hein.journals/hurq21&div=9&id=&page.

Kaeb, C. (2008). Emerging issues of human rights responsibility in extractive and manufacturing industries: Patterns and liabilities risks. *Northwestern Journal of International Human Rights, 6*(2), 327.

London Mining Network. (2015). *European Parliament surprise vote for stronger conflict minerals regulation*. Retrieved May 22, 2015, from http://londonminingnetwork.org/2015/05/european-parliament-surprise-vote-for-stronger-conflict-minerals-regulation/
Murphy, D., & Bendell, J. (1999). *Partners in time? Business, NGOs and sustainable development* (Discussion Paper No. 109. Geneva: UNRISD.
OECD. *Due diligence guidance for responsible supply chains of mineral from conflict-affected and high risk areas*. OECD Guidance. Retrieved April 22, 2015, from http://www.oecd.org/corporate/mne/mining.htm
OECD. (2013). *Upstream implementation of the OECD due diligence guidance for responsible supply chains of minerals from conflict—affected and high—risk areas final report on one—year pilot implementation of the supplement on Tin, Tantalum, and Tungsten*. Retrieved May 22, 2015, from http://www.oecd.org/corporate/mne/DDguidanceTTTpilotUpstreamJan2013.pdf
OECD. (2015). *Responsible gold also means supporting livelihoods of artisanal miners*. The OECD register. Retrieved May 18, 2015, from http://oecdinsights.org/2015/03/24/responsible-gold-also-means-supporting-livelihoods-of-artisanal-miners/
Pedersen, E. R., & Andersen, M. (2006). Safeguarding corporate social responsibility in global supply chains: How codes of conduct are managed in buyer-supplier relationship. *Journal of Public Affairs, 6*(3–4), 228–240.
Public-Private Alliance for Responsible Minerals Trade (PPA). http://www.resolv.org/site-ppa/
Roberts, S. (2003). Supply chain specific? Understanding the patchy success of ethical sourcing initiatives. *Journal Business Ethics, 44*(2–3), 159–170.
Ruggie, J. (2008). *Promotion and protection of all human rights, civil, political, economic, social and cultural rights, including the right to development*. Human Rights Council. Agenda item 3—A/HRC/8/5 7 April 2008—Report of the Special Representative of the Secretary General on the issue of human rights and transnational corporations and other business enterprises.
Ruggie, J. (2013). *Just business*. New York: W.W. Norton.
Sarfaty, G. A. (2015) Shining light on global supply chains 56-2 Harvard International Law Journal 419-463. http://www.harvardilj.org/wp-content/uploads/562Sarfaty.pdf
Scherer, A. G., & Palazzo, G. (2011). The new political role of business in a globalized world: A review of a new perspective on CSR and its Implications for the Firm, Governance, and Democracy. *Journal of Management Studies, 48*(4), 899–931.
Secretary General of the United Nations Consultation. (2007). *Improving the human rights performance of business through multi-stakeholder initiatives: Summary report*. Retrieved May 30, 2015, from http://business-humanrights.org/sites/default/files/media/bhr/files/MSI-Consultation-6-7-Nov-2007.pdf
Stadtler L (2012). *Designing public private partnerships for development*. PhD thesis, University of Geneva.
State Department. (2017). *Notice of stakeholder consultations on responsible conflict mineral sourcing*. Retrieved March 27, 2017, from https://www.federalregister.gov/documents/2017/03/27/2017-05972/notice-of-stakeholder-consultations-on-responsible-conflict-mineral-sourcing
UN Human Rights Council, 17/4 Human rights and transnational corporations and other business enterprises, A/HRC/RES/17/4. https://business-humanrights.org/sites/default/files/media/documents/un-human-rights-councilresolution-re-human-rights-transnational-corps-eng-6-jul-2011.pdf
Utting, P. (2002). *Regulating business via multi-stakeholders initiatives: A preliminary assessment*. Geneva: United Nations Research Institute for Social Development.
Utting, P. (2005). Corporate responsibility and the movement of business. *Development in Practice, 15*(3–4), 375–388.
Zammit, A. (2003). *Development at risk: Rethinking UN-business partnerships*. Geneva: South Centre and United Nations Research Institute for Social Development.

CSR, Innovation and Human Resource Management: The Renaissance of Olivetti's Humanistic Management in Loccioni Group, Italy

Mara Del Baldo

1 Introduction

The chapter addresses the theme of Corporate Social Responsibility (CSR) policies concerning labor and employees by proposing an analysis—developed from both the theoretical and empirical perspectives—of business and leadership models that are geared toward a multidimensional development (economic, social, environmental, ethical) (Sorci 2007; Ketola 2008) which is first and foremost tied to an anthropological and ethical-based perspective (Melé 2009a, b).

CSR refers to the notion of responsibility for the impact of corporate activity on the wider frame of stakeholders, both internal and external stakeholders (employees, customers, banks, suppliers, competitors and social stakeholders, family members, the physical environment, the government, trade and business association, etc.), and it is this attribution of responsibility that underpins the willingness of society to legitimate business (Gray et al. 1996, 2014).

Employees are among the key stakeholders for the development of any CSR strategy or program. The actions addressed to employees are placed among the key internal dimensions of CSR (EU 2002). Their commitment and motivation is fundamental in for companies implementing CSR programs and policies—particularly when companies operate globally in multicultural contexts—as they impact the organizational context, shape employee perceptions, affect employee commitment and consequently motivation. The organizational culture, or "how work gets done around here", is a key dimension of any effective CSR agenda (European Alliance for CSR 2008: 11).

M. Del Baldo (✉)
Department of Economics, Society and Politics, University of Urbino Carlo Bo, Urbino, PU, Italy
e-mail: mara.delbaldo@uniurb.it

H. Lu et al. (eds.), *Building New Bridges Between Business and Society*, CSR, Sustainability, Ethics & Governance, DOI 10.1007/978-3-319-63561-3_10

With specific reference to the European system—which is oriented toward the objective of making Europe a center of excellence in the realm of CSR-, among the internal and external dimensions of CSR, two recommendations in particular address the issue of the firm's employees. Along with ensuring workers' safety and security of employment, emphasis is placed on human resources management policies, which must be oriented toward: ensuring continual formation and instruction throughout the entire lifetime of the employees; pursuing the best equilibrium between work, family, and free time; including employees in the company's management and benefits; informing them of the company's activities, which stimulate their sense of belonging by fostering a shared set of values; enhancing personnel's responsibility; applying the principles of equality in salaries and in the perspective of careers for women; attracting and retaining qualified workers; hiring the disabled, women, and people who are otherwise unemployed. The positive direct and indirect results of these practices are emphasized since they increase the effectiveness of human resource management. Sustainable growth and larger and more effective employment are two challenges to confront when safeguarding the European model of society (EU Commission 2001, 2002, 2006, 2010), which is based on equal opportunities, a high quality of life, social inclusion, social cohesion and a healthy environment. Enterprises, the engine for economic growth, job creation and innovation, are key actors in delivering these objectives, and specific emphasis is put toward the role of SMEs (particularly of family SMEs) in the entrepreneurial fabric and in the socio-economic framework of Europe (IFERA 2003).

As CSR becomes an acknowledged component in employee engagement and therefore the driver of business value, is it expected that CSR alignment will continue to become a critical tool for fostering company success over time (Melcrum 2006). Human Resource Management drives policy development and program implementation in HR areas that can support CSR values (CBSR 2003; CIPD 2002; Koos 2012) and contribute to the establishment of effective CSR policies by delivering on the responsibilities that companies have to their employees as stakeholders (Collier and Esteban 2000; Mees and Bonham 2004; Valentine and Fleishman 2008).

Notwithstanding the growth in the use of codes as a benchmark for employee behavior, there appears to be a widening gap between a company's stated values (mission statements and codes of ethics) and principles and their relevance to organizational practice (Webley and LeJeune 2005; Painter-Morland 2008). Too often in most cases, ethics programs seem to be 'decoupled' from, as opposed to integrated into, every day practices and procedures (Weaver et al. 1999). Existing research emphasize two levels at which effectiveness is determined: firstly the range of organizational practices and procedures designed to create 'an ethical culture' across the organization; secondly the enthusiasm with which employees espouse the organization's attempts to counter the effects of 'negative externalities' by ensuring that their activities create social, environmental and economic capital (Collier and Esteban 2000). High commitment human resource practices and trust in management have a major impact on building employee commitment (McElroy 2001; Whitener 2001). Employee perceptions of justice and fairness are strictly concerned with the ways in which organizations treat employees (Buckley et al. 2001; Weaver 2004). On one hand, if culture is integrative, shared meanings are held in common and there is consensus, consistency and clarity. On the other hand, organizational climate depend, to a large extent, on the quality and stance of management/entrepreneurial team and on the values to which it subscribes (Denison 1996; Treviño et al. 1998).

In the last years there has been a strong re-evaluation on the role of ethically connected values inspired by the managers/entrepreneurs and spread in the company moving toward forms of "humanistic management" (Zamagni 1995; Argandoña 1998, 2003, 2008; Melé 2002, 2009a, 2012) and "humanistic governance" (Pirson and Turnbull 2011), which render businesses as agents of civilization (Röpke 1960; Novak 1996). The frontier of responsible competitiveness is marked by values-driven businesses (Cohen and Pruzak 2001; Cohen and Warwick 2006) in which entrepreneurial success is strictly tied to the capacity to create shared value with stakeholders. "Companies will fail to convince stakeholders that they are serious ("authentic") about CSR unless they can demonstrate that their policies consistently achieve the desired social, environmental and ethical outcomes" (Collier and Esteban 2000: 19). An integrated CSR culture is possible and effective where external-led policies and actions used to support reputation and stakeholder engagement are matched by internal business strategies and decisions driven by social and environmental principles (Hancock 2005).

"Corporate culture flows from and is the consequence of corporate identity (...) the fundamental style, quality, character and personality of an organization, those forces which define, motivate, and embody it, its unique history, business mix, management style, communication policies and practices, nomenclature, competences, and market and competitive distinction" (Downey 1986: 7). The promotion of an ethical culture—which encompasses both culture and climate—depends on how ethics are embedded in the organization (Clarke 2004). "Salient behaviors—awareness of ethical issues, commitment to the organization, integrity, willingness to communicate openly about problems, to seek advice and reduced unethical conduct, improved decision-taking generate employee behaviors" (Collier and Esteban 2000: 26).

Leadership has a crucial role in shaping ethical organizational culture through leaders' moral behavior, corporate mission, vision and values, ethical criteria for recruiting/selection/promotions, evaluation processes and monitoring, ethical training programs, applying ethical values to decision making and in intra-organization procedures and structures. Particularly, personnel policies and practices are a way to bring ethical values into every day routine (Melé 2006). A fundamental element for the effectiveness of ethical commitments and the development of internal and external trust lies in a strong 'steer' from the top (Collier and Esteban 2000: 29–30). Ethical leadership requires personal attributes (honesty, ability to listen, allowing others autonomy of choice, openness, willingness to consult and to learn) and manifests itself in a series of executive behaviors (i.e. gaining a real understanding of the culture; building ethics and values into 'hearts and minds' by means of ethics training programs; building ethical achievement into performance evaluation, creating channels of communication between the company and stakeholder groups, starting with employees) (Schein 1983; Treviño and Brown 2004; Malloch 2009; Gröschl 2013; Capaldi 2013).

Departing from this theoretical background, the following work addresses the attention to humanistic management models of business and leadership, which put specific emphasis on the "flourishing" of employees. The first part synthesizes the theoretical context on which the empirical analysis is found by proposing a brief literature review on the moral-based management and leadership, while the second one analyses the specific business model and policies of CSR geared toward employees that have been successfully implemented by a medium-sized Italian firm—Loccioni Group—which is included among the "great place to work" on the national and international scale. For years, this company has been distinguished for its best CSR-oriented practices regarding HRM, innovation, environment and for its capability to "thread networks" with internal and external stakeholders characterized by a genuine commitment (Zsolnai 2002), which are the result of an authentic and solid value-based system and of a model of exemplary governance aimed at linking economic well-being, social cohesion and environmental protection. A discussion and concluding reflections follow.

2 The Ethical Anchoring of a Good and Virtuous Leader

"When one thinks of a good managerial leader, one thinks about a person who can get the most out of others. An excellent leader is a person who can help others become aware of what they can achieve" (Bertland 2009: 145). Such a leader is able to motivate others to excel and to provide the resources allowing people to develop their capabilities in a way that coheres with communities near and far.

A virtuous manager will need to recognize instances when she/he could help another develop a capability (Nussbaum 2000; Sen 1999). Virtue ethics represents a developing approach within business, resulting from the "recovery" of the idea of virtue in mainstream philosophical ethics in the second half of the twentieth century (Anscombe 1958; MacIntyre 1985; Battaly 2010; Alford and Signori 2014: 5). Among the virtue ethics is a Neo-Aristotelian approach applied to business ethics (Solomon 1992a, b; Hartman 1998, 2011; Melé 2006; Melé 2009a, b). Virtue ethics interprets business to be a practice and attempts to ground ethics on the basis of character rather than rules (Bertland 2009), while emphasizing integrity by suggesting that a person's character needs to be ordered and whole, bringing together home life and work because community is important for fostering virtue (Solomon 1992a, b).

Recently, scholars have begun to stress the relevance of cardinal virtues (fortitude, prudence, temperance, justice; St. Thomas Aquinas) in the business context both at the individual and organizational level (Malloch 2009; Melé 2009a, b; Ruisi et al. 2009; Ruisi 2010; Del Baldo 2013a, b). Fortitude (Courage) means persevering or pursuing what is good in spite of obstacles. Prudence (practical wisdom) aids practical rationality in identifying the right thing to do in each situation and supports the suitable means for the attainment of purposes, ultimately self-realization. Temperance (Moderation) is the ability to control one's emotions by accepting her/his deficiencies. Justice (Friendship) is a virtue that regulates relationships with others (commutative/reciprocal, regulative and legal justice). In a broad sense, justice refers to benevolence and care for the good of others and thus is close to *caritas*, communion and friendship. It comprises all of the virtues regarding human relations including: honesty, loyalty, gratitude, generosity and solidarity.

In a business context, fortitude can be seen as the capacity of adjusting acts to true capabilities and is central for operating decisions, since it determines the coherence of doing with being, thus improving the quality of operating decisions (Bastons 2008). Prudence is the moral competence in predicting, and forms the cognitive structure from which all of the results from decisions are foreseen. It provides the conformity of an action to a real situation (Pieper 1966: 10). Temperance (intended as balance between intemperance and insensitivity) is a quality that enables people to adapt intention according to rational knowledge, trying to impel the action in another direction and is fundamental in decision-making. Other similar moral virtues are generosity—balance between avarice and prodigality, magnanimity—balance between vanity and humility, mildness—balance between quick temperedness and idleness.

"It is incontrovertible that ethics plays an important role in the creation of a business environment in which virtues and values are brought into a relationship for the good of all. In this regard, character and, in particular, the character of leaders is paramount" (Flynn 2008: 360). The same author suggests that a new concern for the integral needs of a person (psychological, social, cultural and spiritual) with the aid of the owners and managers of the business, would help to reduce some of the most deleterious trends in modern society. Personal responsibility in business requires imagination, creativity, and financial resources). The ethical-bases and virtue construct supplies a conceptual (and managerial) framework for a CSR authentic strategy (Del Baldo 2013a, 2017). In this sense, the virtue matrix proposed by Martin (2002) addresses questions about corporate responsibility and *is* aimed to help executives understand what generates socially responsible corporate conduct.

3 The Moral-Based Leadership and Management

There are leaders who freely admit that they are driven by a more intrinsic and contagious commitment to values (Bouckaert 2011). Hoivik clearly underlines how and why leadership is not possible without ethics and how one cannot separate them, as "being a moral leader and doing, acting with moral leadership are one" (Hoivik 2014: 4) as shown in Table 1.

Morals and leadership can be studied on an individual level and on a group and organizational level (Bass and Bass 2008). Moral behavior[1] and leadership are

[1]Moral behavior is defined as the ability to implement justice requirements derived from a fair distribution of rights and duties in a demanding operational context (Greenberg and Colquitt 2005).

Table 1 The moral leader: being and doing

Being informed by values, emotional and reasoning capability, caring, visionary, proactive and innovative	**Doing** informed by relationship with all stakeholders, wanting to achieve the best for all and the common good
• Balanced/in harmony with yourself • Reason & emotions • Integer (integrity) • Vision • Passionate • Responsible (ethical) • Trustworthy	• Relational • Affective • Being there (crisis) • Caring • Communicating • Involving others • Responsive

Source: our adaptation of Hoivik (2014)

interrelated: on one hand, moral behavior is influenced by situational factors such as role modeling, diffusion of responsibility, conformity, etc. (Zimbardo 2007); on the other hand individual differences (i.e. personality and values) act as antecedent of moral behavior (Brown and Treviño 2006; Rest and Narvaez 1994).

The organizational culture imbued with moral leadership enjoys several benefits: understanding of the interdependence with stakeholders; learning environment; respect and trust; cooperation; responsibility and accountability. As one can see in the following sections, this is not an idealistic approach, being that it is possible to find examples (i.e., the Norwegian clothing company Stormberg A/S; see Hoivik and Melé 2009) of organizations that apply it, including where moral leadership is inspired by constitutive moral elements: innovation, intuition and imagination (Hoivik 2014). Moral imagination entails perceiving norms, social roles, and relationships entwined in managerial decision making; moreover it involves the ability to envision and evaluate new models that create new possibilities to reframe problems and create new solutions in ways that are economically viable and morally justifiable (Werhane 1999: 93). Moral leadership and moral creativity have become more important than ever for businesses that have to face a global environment because neither philanthropy nor risk management are sufficient any longer. Creative value management depends on the attention paid to all values that are at stake.

Moral leadership is considered a key driver to implementing authentically CSR and sustainability-driven strategies. CSR requires moral creativity to foster a greater normative and creative approach toward CSR and sustainability strategies and tools, which should be dynamic and innovative. The development of CSR and sustainability-oriented leadership are considered as an ongoing process, requiring both creativity and moral conviction (Visser 2011). Sustainability leadership entails a focus on all management actions and activities of the economic, environmental and social objectives of a company, as well as on the interdependencies between them. Running a company that is genuinely oriented toward sustainability involves more than merely implementing a separate program "in addition" to existing conventional processes (Von Ahsen 2015; Oreg and Berson 2011). It requires exercising a leadership based on values (such as caring, people-centredness and integrity) that direct the corporation so as to ensure its prosperity based on transparency, accountability and responsibility.

It is also critical that leaders set responsible examples, so that followers imitate them in the process of carrying out their duties. The ability to "walk the talk of morals" is emphasized (Bass and Steidlmeier 1999; Simons 1999) when maintaining trustworthiness and model attractiveness (Choi and Mai-Dalton 1998) and is positively related to charismatic and transformational leadership (Choi 2006; Liu 2007; Olsen 2010; Brown 2011; Palshikar 2007).

Charismatic leaders have turned problems into opportunities and resources thanks to their different ability "to see the world" (Bruni and Sena 2013), and are able to create and maintain a work environment where people are emotionally and intellectually committed to the organization's goals. They build an energetic and positive attitude in others and inspire them to do their very best by creating a common sense of purpose (Jacobsen 2001; Palshikar 2007; Opdebeeck 2013). Personal responsibility, vision, moral virtues, integrity, faith in personal commitment, shared social responsibility, and solidarity are typical charismatic leaders' attributes (Becker 1998) that leverage a virtuous corporate culture in an organization. The research Globe (global leadership and organizational behavior effectiveness) (2008)—aimed at understanding the cultural characteristics that positively or negatively influence the leader effectiveness—identified several leadership dimensions among which are ranked: charisma and ability to motivate members of the organization by leveraging the transmission of corporate values; the ability to create and manage working groups and orient them toward common goals; the level of members involvement in the decision-making process; the level of compassion generosity and the ability to provide human support to the members of the organization.

The transformation of followers' moral values are appreciated as an important approach in order to bolster commitment toward the common good, even if it is noted that this can be a somewhat idealistic approach that should be supplemented by mutually binding agreements related to the distribution of duties and rights (Bass and Steidlmeier 1999), as emphasized by Brown and Treviño (2006), who sees the establishment and maintenance of (moral) rules and agreements as a necessary aspect of ethical leadership and cooperation.

4 The Loccioni Group: The Business and Leadership Model

4.1 *Methodology*

The empirical study was developed according to a qualitative approach and a case study methodology (Eisenhardt 1989; Yin 1994). Recently scholars have called for a return to in-depth methods, such as narrative and case studies (Gartner 2007), which are valuable when generating theoretical propositions (Eisenhardt and Graebner 2007).

Following this approach, the study focuses on the Loccioni Group (single case study), an Italian enterprise that is authentically CSR and sustainability-oriented (Del Baldo 2013a, b; Del Baldo 2017). The Loccioni Group is medium-sized, not listed on the stock market and is family run. It is therefore a typical example of the "family businesses" which is extremely widespread in the Marches Region. This region is emblematic of the "Third Italy" model in which the development of SMEs is established in small towns, without disrupting the pre-existent agricultural and craftsman vocations and also preserving the socio-economic fabric of relationships anchored in the territory. Since Italy has the largest presence of artisan companies and districts, the provinces within the Marches regions come out on top in the national list for balancing economic development with social cohesion, and also for the diffusion of "best practices" companies for their CSR and sustainability-oriented development projects listed by ISVI (the Institute for Business Values), that have also been recognized at the national and international levels (i.e. the Sodalitas Social Award). (Unioncamere 2010, Marchegian excellent companies—Istao 2014).

The analysis was based on information collected from 2013 and is still in progress through in-depth semi-structured interviews, informal conversation with the founder (Loccioni Enrico), his entrepreneurial family and with other collaborators at various levels and firm-wide functions, as well as local partners. Moreover data have been collected through direct and participant observation during companies' visits, conferences and focus groups, in which entrepreneurs' experiences were exchanged with researchers, other entrepreneurs, and local institutions—and through documentary analysis (Bartocci 2011; Varvelli and Varvelli 2014). In total, there were more than 20 meetings during which interviews, workshops and focus groups were carried out, transcribed and validated with the interlocutors. Some summaries of interviews will be included in the presentation of data and the discussion of results.

4.2 Company Profile

The Loccioni Group was created in 1968 in a small town near Ancona (Italy) by Enrico Loccioni (the current president), who founded six businesses in the next 40 years. The Group, which is formed by six businesses counts more than 350 employees and a total sales (mainly international) of over 50,000,000 Euros (2014).

Ownership is primarily familial. Loccioni Group's activities are based on distinctive "core competencies" aimed at custom-design innovative technological solutions for its clients and applied to diverse markets: environment, home, automotive, industrial, community, health: integrated technologies for environmental monitoring; measurement and quality control; biomedicine and medical equipments; telecommunication and environmental control; green energy; training and consultancy for technical, managerial education and for business development.

The Loccioni Group adopts processes of social and environmental certification, regularly publishes social and environmental reports, and has obtained recognitions for its robust activities of social responsibility and sustainability both on the national and international level. Among the tools for implementing and communicating CSR and sustainability we can mention: the list of company values (since 1969); the Code of ethics (since 1996); the social report (since 1997); intangibles impact (since 1997); cause related marketing (since 1999).

There are four distinctive traits of the group: (1) a solid ethical base of the entrepreneur—which is nurtured by religious principles and values inherited from his agricultural forefathers—that is transmitted by the founder to the entire organization; (2) an outstanding capacity to nurture relationships and communication which leads to an organizational model founded on CSR-oriented networks and partnerships and on an intense activity of a stakeholders dialogue; (3) a strong sense of belonging to the local community and a sincere love for its own Marchegian "land", perceived as a stakeholder and an authentic willingness to ameliorate the environment, beginning with the local level and (4) an authentic "faith" in human beings and the qualities (talents) that every person may improve on throughout his/her work and private life.

Numerous awards have been attributed to Loccioni Group for its CSR and sustainability-oriented projects including for its excellent level of innovation. One can cite for instance the Sodalitas Social Awards: finalist in 2005 for the Internal Processes of CSR and network enterprise model "Metalmezzadro" in the knowledge-based business; in 2008 for its "Sustainability Projects" and in 2009 for the Project LOV (The Land Of Values). Moreover the group obtained the "Business and Culture" Award (2003) for the project "Bluzone"; recognition from Legambiente (2005) for being a partner of the European Commission in the "Sustainable Energy Europe Campaign" with the "Leaf Community Project Leaf Energy and Future", the Best Workplaces Italia (awarded from 2002 to 2007) and the "Great Place to Work" (awarded in 2014).

At the same time, numerous are the projects which are distinctive features of the organizational model of the group, whose mission is "with curiosity and openness we integrate ideas, people and technologies to animate and give value to the business"—based on internal and external networks: U-net, a multidisciplinary network of universities and research centers for the development of scientific competence and applied research; Crossworlds, a network of international groups aimed to stimulate automotive-based knowledge and know-how among other sectors; "Bluzone", an educational laboratory (accredited since 2002 by the Marches Region) that partners with more than 28 schools, 20 universities and 5 Master programs (every year more than 1000 students are hosted); Nexus, a multi-sector network of local businesses—which group more than 30 entrepreneurs and over 550 collaborators—created in 1994 by Enrico Loccioni's idea to facilitate local SMEs' inter-firm collaboration and to increase territorial growth through sharing knowledge and experiences. The initiatives promoted by Nexus include monthly meetings, virtual board meetings, training courses, partnerships with schools, scholarships, providing data for students' thesis, polytechnic visits, and European projects.

4.3 The Strength of Loccioni Group's Values

The entrepreneurial vocation that makes the Loccioni Group's excellence derives from a rich set of values (trust, respect, gratitude, humility, communication, willingness to sacrifice, concentration, determination, respect for labor and work ethics) spread from the founder's—Enrico Loccioni—leadership model which is shared by the entire organization.

Values that guide the decision-making process in adherence to an authentic CSR-oriented vision are anchored to virtues which are testified first by the founder—who is distinguished by his charismatic personality and his ability to interpret reality through a different outlook by transforming problems into opportunities for the community and for the business—and then diffused among the entrepreneurial family and the group's employees.

The first chart of shared values was adopted by the Loccioni Group in 1992 and included the following principles (Table 2).

In 2005, the Loccioni Group set its higher principles that integrate the values previously stated (Table 3).

> With these values we try to create value for customers, for the company, for the people and the community everyday (E. Loccioni—ID 20).

> We look for people who share our values and ideals consistent with our way of imagining the future and the world around us. It will be the daily approach to working that will confirm the people who have the habit of virtues, and to distinguish them from those who simply "wear the dress" (Loccioni Group—I. Terzoni, 27/05/2014).

Loccioni Group's values are not only related to individuals but to the entire organization; they are not abstract concepts but are translated into virtues through implementing good behaviors (Table 4).

4.4 The Renewed "Olivettiano" Humanistic Management in the Loccioni Group

The Loccioni Group is conceived and exists as a "play factory," that is in an enthusiastic, passionated and creative way (Bartocci 2011). It is a knowledge enterprise in which there are not employees but collaborators, and in which projects are developed and not products. The business is experienced as a continuous adventure (from the Latin "ad-venire") , that is a company that creates what will take place "ad-venire" (E. Loccioni. See: Bartocci 2011).

The peculiar business model envisioned by the founder (E. Loccioni) is the result of a model of leadership able to develop a shared culture, which has role models such as Merloni (the Marchegian founder of the Merloni Group), E. Mattei (ENI), W. Von Siemens, and particularly Camillo and Adriano Olivetti.

Table 2 Loccioni Groups' milestones

Initiative and Intelligence	Initiative and intelligence are associated with prudence, that is to be "entrepreneurs" and be able to choose the opportunity, internalize corporate values, by acting quickly and taking responsibility for one's own actions Prudence allows them to carefully evaluate each situation, consider the appropriate means to achieve the end result, and therefore consists in having the intelligence to choose the opportunities as a result of a clear and comprehensive analysis of the whole context
Energy and will	"Anything is possible if every daily action is based on commitment, sacrifice, tenacity and perseverance (if one really wants it)" Energy and will are associated respectively to strength and justice: justice is the will of giving to each his right; strength instead results from a constant practice of knowing how to direct energies toward solving problems, understanding the situation, without being discouraged by the possible obstacles in which one faces
Flexibility and adaptability	It is the orientation to change, which does not impose solutions but rather searches for solutions together with those that present a problem Flexibility and adaptability are associated to strength or fortitude, that is the ability to accept one's own limits, the possible falls and vulnerability. It is derived from a high propensity to change and the capacity to constantly look for new solutions It is the ability to accept changes, to face adversity, and even deal with situations of "failure" perceiving them as an opportunity for change
Innovating one's self to innovate	To find new solutions needed to change their own beliefs, attitudes and behaviors Innovating one's self to innovate is associated with the fortitude and temperance: it involves constantly questioning situations and changing their own beliefs, attitudes and behavior. Practicing this value can reach fortitude, because innovation spurs change; consequently the risk (healthy and controlled risk) and drive to grasp more entrepreneurial opportunities over time generates temperance, i.e. ability to stay balanced in times of change, without exceeding enthusiasms, discomforts or insecurities
Transparency in communication	Entering into relationship with frankly everyone, at all levels, devoting time and energy to understand and be understood Transparency in communication is associated to justice: that value implies honesty and openness of collaborators, and cultivates over time, guaranteeing justice in one's work
Listen to anticipate	Listen especially to the "weak noises" which, if not heard in time, can become destructive discordance Listen to anticipate is associated with the virtues of temperance and prudence. In fact, the temperance and moderation of instincts requires attention, and listening is the practice that takes time and attention to be done well. Listening spurs human beings waiting for their turn to speak, to the analysis of the situation and the interlocutor, thus even generating prudence

Source: our elaboration of Loccioni's "Dodecalogo" (list of values)

Table 3 The new chart of values

List of Loccioni Group's values	Content
Energy: Much of it is needed to dream and to realize one's dreams	It means putting enthusiasm, passion, courage and motivation in all that we do. It is the smile, kindness, joy, and tenacity
Responsibility: For the air that we breathe, the land that we walk, the resources that we utilize, and the trust that we obtain	It means taking charge of the future, being aware that every action and project has consequences. And it is the answer to the trust offered to us (by customers, suppliers and collaborators). It is the ethical dimension of being a person in a community
Imagination: Being capable to imagine means being capable to create	It is the ability to dream, "to look with open eyes," to see the invisible, to ask useful questions to fulfill the envisioned dream. It is the desire to participate in building the future
Tradition & Innovation ("Tradinnovation"): Learning from the past to give form to the future	It is the link to tradition, listening and using the experience to look to the future and innovate. This means not forgetting where we come from, able to recognize our dreams and achieve them. It is "pulling the arrow further back to go farther"

Source: Loccioni's chart of values

Table 4 Loccioni Group's organizational behaviors

Confidence in the team, proxy and accountability
Fault tolerance, available to the exercise, continuous improvement
Ideas, intangible knowledge
International dimension, travels and trips
Willingness to change
Capability to watch over others ("to look over the 'own garden' ")
Proactivity, initiative
Sobriety (no excessive behaviors)
Enthusiasm and shared passion
Networks
The task of sowing beauty (the 'undertaking to sow beauty')
Projects are always different; there is no mass production
Going beyond pure business, without ever forgetting the territory
Courage, initiative, resourcefulness to be "intra-preneurs"

Source: Our elaboration from Loccioni Group data (May 2014, 27th)

The Olivetti group (the first in Italy to produce typewriters and subsequently computers) became a symbol of the progress after World War II, which is the Italian style and the Italian design in the world. In 1956, the company employed more than 16,000 people in different businesses locations in Italy and abroad (Germany and America), with over 100 patents and 300 models. Olivetti was an example of an entrepreneur who aimed to value people both as employees and men because "what is important is to build men, forge character without which education and culture are vain education and culture". This moral imperative is summarized in the following words: "This dual struggle in the material and the spiritual sphere, for this factory we love, it is the highest commitment and the very reason for my life. I hope I have never ignored your aspirations, your desires, your needs. Your hopes have always been mine. For years in prayer, every day I always thought about my daily bread and yours, and about my great responsibility to ensure my future, your future and that of your families". (A. Olivetti, Ivrea, Italy, 1955, 24th December)[2].

These words summarized the sense of a true corporate social responsibility and stakeholder engagement that Olivetti has always implemented through concrete actions, business and social projects (vocational training center for mechanics, canteen, infirmary, houses for employees), based on his own employees and collaborators. This way of "being and doing" business has deeply influenced the Italian social and cultural context and is proof of how it was (and is) possible to think of a business model that is very real, based on the following principles: the close relationship with the university and the entrepreneur scientific culture; the continuous contact with the most innovative foreign companies; the enhancement of the contribution of loyal and tenacious employees; the enhancement of the ability to craft and use creativity, and aversion to repetitive and prolonged work; the factory conceived as a school laboratory, which helps build the knowledge society; an effective leadership, based on the profound knowledge of men and things, and the ability to obtain them without control; the employment relationship based on the principle of equality between men (work made by men who are equal); the commitment to build a true democracy within the community to which they belong; the desire to produce a lot of great products; a governance centered on people of high moral standing, aimed at searching for results in the long term, free from a speculative mentality; the idea of social enterprise as a space capable of recognizing the legitimate aspirations of each person (See: Loccioni Group 2008a: 23).

Loccioni Enrico (the founder) has launched a series of publications dedicated to the commitment of the "true entrepreneur" and the "true value of the firm (factory)" as a place of creation of material and human wealth and as a community of people united by a common mission. Olivetti has often been considered a visionary for having always cultivated the dream of a company that has a social role and is a key player in the socio-economic and cultural context. He was also defined a dangerous example by those who behave in the opposite way and do not intend to change. However, many entrepreneurs like Loccioni Enrico continue to follow and enrich this dream: "There are so many buds and plants in Italy, even if a bit 'hidden and little known'"(Bruno Lamborghini, President of the Association Archivio Storico Olivetti—Olivetti Historical Archive; See Loccioni Group *Cultura d'Impresa*, Vol. 1: 8). Comparing these principles, it is possible to identify the affinity for the Loccioni way of seeing and doing business in the words of the company's leader and collaborators.

[2]See: Loccioni Group, *Cultura d'Impresa*, 2008, Vol. 1 and 4.

> We want to spread a new work culture based on passion, enjoyment and beauty because it is through play that the most intelligence is expressed. Play is at the basis of charisma. Our young people are passionate about doing their job well. Seeing them with a smile in their eyes, seeing their respect and sense of fair play in the work place is something which fills my heart with joy and affection. Everyone of us brings something of our lives to our work and gives something to others just like in a family. We share a simple style of life and agree on the need to invest earnings into the company. There must be a passion for continuous improvement because as a friend of mine, a village priest, said: we are all users of this land and the best thing we can do is leave things in a better way than how we found them, as we have received more than we have given. (E. Loccioni—ID 23).

> Since the beginning of this adventure, the objective, the dream at the base of everything, has been to be the actors in a context of continuous change, to be the creators of the future we want rather than just mere spectators. I have had a single objective which has always been clear since the beginning: don't make others do what I would not have liked, that is, a repetitive job in which everything is decided by the others. I have had and have the dream to create a model of a company aware of its social role and of its future in the territory and the world. To nurture that dream, people must understand the future they want for themselves and for their group. They must find their vision and the managers must help them in this process. (E. Loccioni; See: Bartocci 2011).

From these statements an emerging key concept is a visionary leadership style, like in the mystic religious tradition founded on transcendental values (virtues). Nevertheless, the founder's leadership style cannot be considered idealistic and visionary, because E. Loccioni (as his son who is part of the management group) is a "down-to-earth" man. In particular, like for Olivetti, the enterprise is conceived as a mean to improve the quality of life in that specific territory, where in the past, for a long time, emigration was the scourge of these lands and where there were few (like E. Loccioni) who, instead of leaving the fields, chose to remain in the area creating work and generating better living conditions through the company.

4.5 Loccioni Group: A 'Firm' Based on People

A distinctive aspect of the Loccioni's leadership and business model is the centrality attributed to people. All of Loccioni Group's actions and strategies are not based on employees, but on collaborators, that is those people who work together, sharing knowledge competences and capabilities to improve the company, which is conceived as a common good and as a means for the common good. Relationships are not based on hierarchies (which "imprison" knowledge and intelligence) but on trust, not hierarchical.

> I read about the 'centrality of the person' in almost all the ethical business manuals that I've ever held (I haven't read them all: it seems as though there are ten of thousands of them, even Enron, had them, and which was even awarded...); I heard them being authoritatively and peremptorily discussed by prominent business leaders. But, to tell the truth, in many cases I had a great deal of difficulty in recognizing, in real life and everyday behaviors of leaders and managers, the solemn statements carved on the manuals. Even in the brochures of this company, they discuss the "centrality of the person." But it is a cliché that

corresponds to reality; for example, the centrality of the training (eight thousand hours per year in a company with less than four hundred employees); the care of the working environment; the respect and cordiality of relations (E. Loccioni. See: Bartocci 2011).

Here there are no leaders and we do not even like the word leader. Here the leaders are called responsible people: they must have the ability to listen, communicate, motivate people and give them a chance to rejoice and give meaning to their every day work. No one—including myself—is the head of anyone. In the Loccioni Group, there are people able to pull the others to face difficult challenges with enthusiasm. There are the teachers, who, through their experience can teach both the young, and vice versa. There are coaches, who are willing to take responsibility to raise a team. There are explorers, who jump with a passion for innovative projects involving the team. There are visionaries, those who have the ability to imagine the Loccioni Group as different and a cut above, committed to transferring, sharing, and helping other to fall in love with this dream (E. Loccioni, ID 24).

In 2011, a talent tutoring was launched, based on the voluntary participation of employees. The project shifted the focus of training from management action to entrepreneurship, aimed at finding new ways to build the future. The founder (E. Loccioni) started the project by sending a personalized letters addressed to "those who think they have an untapped talent, who feel like they need to do more and or something else, and who want to prove themselves and would be willing to change."

Here talented people are knowledge players and power players of intrapreneurship. The path to intrapreneurship is based on the following steps: know, do you know, to act (to do), to value and communicate. And it is through the growth of others that one can experience a personal growth. Intrapreneurs are people with the ability to develop the business within the enterprise. The intrapreneur-manager is a person who, through his moral-oriented behaviors, disseminates behaviors and is a manager of culture and civilization. In contrast, there is the butler. We believe in people and their desire "to do" because in their hands, the possibility resides to gain a success that is a common heritage for all. In our region, (the Marches, a million and a half inhabitants) 10% are entrepreneurs! (E. Loccioni, ID 23.)

In our group, optimism, passion and enthusiasm are winning qualities. Coherently to this approach, that can be a problem is experienced with obstinacy, humility and the joy that accompanies the act of creation. This approach is a constructive habit for one's own good and the common good, being and has its strength in its anchorage to virtues. The result is a strong spirit of cooperation from the whole community. The matrix organizational model adopted by the group is dynamic and fluid, and is centered on projects. The project dimension is always inter-functional and gives us a chance to prove ourselves as entrepreneurs and creators of new worlds. (Palermi M.P., External Relationship Manager, ID 32).

The type of work and the level or financial reward does not matter, what really counts is the feeling made by creation, the responsibility to feel useful, the self, the sense of participation that the work returns. No leisure fun than work lived with interest, done in a responsible manner, with improved commitment. (Libenzi R., General Manager—ID 51).

The work ethic is based on the enthusiasm that opens a virtuous circle. The enthusiastic person is strong, positive, and influences those around, impacting them with the qualities of a natural leader. (E. Loccioni, ID 22).

Key concepts that emerge from these statements are the peculiarities of the business model, centered on the people at the base of human capital (Melé 2009a),

Table 5 The organizational climate and well-being in the workplace in Loccioni Group

Code number	Factor	Assessment made by the employees		Gram sample consisting of 183 Italian companies
		Numerical score	Qualitative score	
1	Interest in work	4.6	Excellent	3.1
2	Physical working environment	4.8	Excellent	3.4
3	Flux and fluidity of work	2.3	Scarse	2.9
4	Information about and for the work	3.8	Good	2.5
5	Functional relationships	2.8	Scarse	2.2
6	Interpersonal relationships	4.1	Excellent	4.0
7	Relations with the hierarchy	4.5	Excellent	2.6
8	Company's image	4.9	Excellent	3.7
Total		31.8		24.4

Source: GPTW Loccioni (2013)

the capacity and the will to develop the talents through a transformational leadership (Bass and Riggio 2006; Liu 2007; Oreg and Berson 2011) that is geared toward creating a supportive environment and renders possible the maximization of creativity. The Loccioni Group is distinguished by its innovative high tension, strong entrepreneurial orientation of all employees as well as the ability to recognize and create business opportunities. All of these qualities contribute to flourish the talents of all human beings and, at the same time, the competitive success of the company which, over the years and even in periods of crises, developed on an international scale. Moreover key concepts highlighted by these assumptions and from the analysis are: values related to the virtue of foresight, the spirit of sacrifice, and enthusiasm. These are flanked by the organizational well-being of employees considered as a priority and who can be associated with the virtue of justice strongly testified and developed by the founder.

The Loccioni Group has in fact for years been one of the few Italian companies included among the "Best Workplaces Italia" and "Great Place to Work" Award, along with companies like Ferrari, Coca-Cola, and Microsoft ("Great Place to Work Award" 2014). The first application of the analysis of the organizational climate, then called welfare organization, dates back to 1987. In the early 2000s, the analysis was carried out by an external body (Great Place to Work Institute) in order to objectively and subjectively evaluate the organizational climate of the company, and to cancel the distance between being and seeming to be (Table 5).

The Loccioni's unusual environment is distant from the stereotype of the dark and noisy factory: "This is a place where the arms of the most lowly operator are always an active extension of the brain, and where each product is a creation rather than a chore" (E. Loccioni: See: Bartocci 2011: 68).

5 Discussion and Conclusion

The case study offers an example of the best stakeholders' and employees' management practices, which co-evolves with the environment, improving, at the same time, the company's competitiveness and the socio-economic conditions of the local context in which it is deeply embedded. In this context, CSR is part of the DNA and widespread in the entire organization (Walker and Schmidpeter 2015; Weidinger et al. 2013).

Loccioni Group is an "extreme case" (but not unique in Italy), helping underline the importance of embracing the cultural and anthropological roots of CSR, which adhere to a model of humanistic management that conceives the business as a tool for promoting social, economic, moral and environmental well-being. leadership model—which is closely connected to the leadership model—the creation of corporate shared value is considered to be a new way of "reinventing capitalism". The Loccioni Group represents a diffused entrepreneurial network, a "value retainer" and "cultural or heritage-driven," whose identity is market by basic principles: imagination, energy, responsibility, tradition & innovation (to learn from the past to give form to the future). The Group had envisioned its own model of holistic development in Adriano and Camillo Olivetti and has "reinvented" and reinterpreted the Olivetti's model of industrial humanism, which involves creating economic value in a way that also creates value for society, by addressing its needs and challenges and becoming an educational laboratory and vector of intangible factors to which in the last several years companies, universities and research centers increasingly have turned their attention (Loccioni Group 2008a,b,c).

"The Loccioni Group, thus represents a "case" not so much as an original business experience but rather as an exceptional story of humanity in which it is quite impossible to separate "the reason of the heart" from those of the mind" (Bartocci 2011: 63). From its very beginning, Loccioni Group has had a clearly formulated value based mission statement: wanting to make the world a better place. The values shared in the company are driven toward virtues such us honesty, courage, and responsibility. The principles of innovation, imagination and responsibility (Hoivik 2014) are key drivers of every choice. Collaborators and leaders in such an environment nurture a culture of seeing the other and listening. Involvement and teamwork are common features, and even learning from your mistakes is encouraged. The resulting organizational energy is creating a further development of individuals together with the respective organization.

Talking with the founder, one arrives at the conclusion that he sincerely believes in people and in their ability to contribute with their own unique resources, if only given the opportunity. The sense of caring, as well as having the courage to do things in a simple, yet different way, is fundamental to his notion of understanding ethical leadership and social responsibility. For Loccioni "social responsibility" in companies is fundamentally simple. It is about "'caring about': people, environment and society" (Hoivik and Melé 2009).

Loccioni's view is in line with the Marchegian traditional understanding that business organizations need to engage in the development of the society in which they want to do business (Del Baldo 2014).

Loccioni Groups' moral leadership approach has been developed by its charismatic founder, Mr. Enrico, who has been (and still is) able to pass on his passion and motivation to his collaborators by creating an organizational value based identity. He has been capable to embed his motivation into the company and the employees by caring for employees and adopting a participatory decision making processes, thus sharing responsibility with them. Thus, the company has both a top-down and bottom-up management approach with a focus on a consensus. In addition, the emphasis is on caring for all relevant stakeholders and involving them actively in the direction and policies of the firm, similar to the Judeo-Christian injunction, "To love one another as oneself." The Loccioni Group can be considered a school of virtues (Ruisi 2010), which is a place where one can relate to others and obtain enrichment through virtuous relationships.

The final message that emerges from the testimony of the case is that the company must offer itself as a laboratory where the minds and hearts of those involved have aroused, a good place in which to live a good life (Melé 2002). To this end, it is necessary that companies have "cultured" men, intended as people, who have developed and made their knowledge and know-how, but also values which explain the sunken reasons for action and virtues. In the Loccioni Group, there is a strong hope that the company is engaged in a training experience, in a sort of continuous adventure, and in a challenge. Therefore, this awareness is linked to courage (courage of the decisions and the consequent responsibility), patience (to face events and unexpected behavior, and the capability of awaiting unforeseen events and good results) and fortitude. The ethical and virtues-based entrepreneur is thus related to the repetition of good acts aimed at meeting the expectations of stakeholders and to balance long-term development of the company.

The focus on a renewed entrepreneurial ethos first and foremost nurtures the virtues, rather than information and operational/managerial capabilities, and is fundamental for an authentic entrepreneur or manager, oriented to the true and good, and interested in the beauty, and the ability to respect and value his dignity and that of every human being.

References

Alford, H., & Signori, S. (2014). Brief considerations on the effectiveness of shareholder activism. A virtue ethics approach. *Impresa Progetto-Electronic Journal of Management, 3*, 1–10.

Anscombe, G. E. M. (1958). Modern moral philosophy. *Philosophy, 33*, 1–19.

Argandoña, A. (1998). The stakeholder theory and the common good. *Journal of Business Ethics, 17*(9), 1093–1102.

Argandoña, A. (2003). Fostering values in organizations. *Journal of Business Ethics, 45*, 15–28.

Argandoña, A. (2008). Anthropological and ethical foundations of organization theory. In S. Gregg & J. R. Stoner Jr. (Eds.), *Rethinking business management. Examining the foundations of business education* (pp. 22–37). Princeton, NJ: Witherspoon Institute.

Bartocci, M. (2011). *Animal spirits in Vallesina. Enrico Loccioni e "l'impresa come gioco"*. Milano: Luiss University.

Bass, B. M., & Bass, R. (2008). *The bass handbook of leadership* (4th ed.). New York: Free Press.

Bass, B. M., & Riggio, R. E. (2006). *Transformational leadership* (2nd ed.). Mahwah, NJ: Lawrence Erlbaum.

Bass, B. M., & Steidlmeier, P. (1999). Ethics, character, and authentic transformational leadership behavior. *Leadership Quarterly, 10*(2), 181–217.

Bastons, M. (2008). The role of virtues in the framing of decisions. *Journal of Business Ethics, 78*, 389–400.

Battaly, H. (2010). Introduction. Virtue and vice. *Metaphilosophy, 41*, 1–21.

Becker, T. (1998). Integrity in organizations: Beyond honesty and conscientiousness. *Academy of Management Review, 23*(1), 154–161.

Bertland, A. (2009). Virtue ethics in business and the capabilities approach. *Journal of Business Ethics, 84*, 25–32.

Bouckaert, L. (2011). Spirituality and economic democracy. In L. Zsolani (Ed.), *Spirituality and ethics in management*. Dordrecht: Springer.

Brown, J. B. (2011). The building of a virtuous transformational leader. *The Journal of Virtues and Leadership, 2*(1), 6–14.

Brown, M. E., & Treviño, L. K. (2006). Ethical leadership: A review and future directions. *The Leadership Quarterly, 17*, 596–616.

Bruni, L., & Sena, B. (Eds.). (2013). *The charismatic principle in social life*. New York: Routledge.

Buckley, M. R., Beu, D. S., Frink, D. D., Howard, J. L., Berkson, H., Mobbs, T. A., & Ferris, G. R. (2001). Ethical issues in human resources systems. *Human Resource Management Review, 11* (1), 11–29.

Capaldi, N. (2013). How American spiritual capital inform business and affects the common good. In S. Gröschl (Ed.), *Uncertainty, diversity and the common good. Changing norms and new leadership paradigms* (pp. 25–40). Surrey, England: Gower.

CBSR. (2003). *Best people practices: The integration of values into people practices*. Vancouver, BC: Author.

Choi, J. (2006). A motivational theory of charismatic leadership: Envisioning, empathy and empowerment. *Journal of Leadership & Organizational Studies, 13*(1), 24–43.

Choi, Y., & Mai-Dalton, R. R. (1998). On the leadership function of self-sacrifice. *Leadership Quarterly, 9*, 475–501.

CIPD—Chartered Institute of Personnel and Development. (2002). *Corporate social responsibility and HR's role*. London: Author.

Clarke, M. M. (2004, September 28). New sentencing guidelines to reward ethical culture, compliance, commitment. *HR Magazine*.

Cohen, D., & Prusak, L. (2001). *In good company: How social capital makes organizations works*. Cambridge, MA: Harvard Business School Press.

Cohen, B., & Warwick, M. (2006). *Values-driven business: How to change the world, make money and have fun*. San Francisco: Berret-Koehler Publishers.

Collier, J., & Esteban, R. (2000). Systemic leadership: Ethical and effective. *The Leadership and Organizational Development Journal, 21*(4), 207–215.

Del Baldo, M. (2013a). CSR-oriented SMEs: A question of entrepreneurial virtues in action? Reflections in theory and practice. In J. O. Okpara & S. O. Idowu (Eds.), *Corporate social responsibility. Challenges, opportunities and strategies for 21st century leaders* (pp. 145–170). Berlin: Springer.

Del Baldo, M. (2013b). Stakeholders management approach in Italian 'territorial' companies. Loccioni Group and the 'Land of Values—LOV' Project. *European Journal of International Management, 7*(2), 225–246.

Del Baldo, M. (2014). To the roots of entrepreneurial values: The relationship with the territory as a driver for the development of corporate social responsibility—An analysis of the experiences of Italian SMEs. In A. Yüksel Mermod & S. O. Idowu (Eds.), *Corporate social responsibility in the global business world* (pp. 169–189). Berlin: Springer.

Del Baldo, M. (2017). Authentic CSR and Leadership: Toward a virtues-based model of stakeholders dialogue and engagement. The Loccioni Group experience. In S. Vertigas & S. O. Idowu (Eds.), *Stages of corporate social responsibility: From ideas to impacts*. Berlin: Springer.

Denison, D. R. (1996). What is the difference between organizational culture and organizational climate? A native's point of view on a decade of paradigm wars. *Academy of Management Review, 21*(3), 619–654.

Downey, S. (1986). The relationship between corporate culture and corporate identity. *Public Relations Quarterly, 31*(4), 7–13.

Eisenhardt, K. M. (1989). Building theories from case study research. *Academy of Management Review, 14*(4), 532–550.

Eisenhardt, K. M., & Graebner, M. E. (2007). Theory building from cases: Opportunities and challenges. *Academy of Management Journal, 50*(1), 25–32.

EU. (2001). Green paper *Promoting a European framework for corporate social responsibility*, COM(2001) 366 final. Bruxelles.

EU. (2002). *Corporate social responsibility: A business contribution to sustainable development*, Com(2002)347 final, Bruxelles.

EU. (2006). *Implementing the partnership for growth and jobs: Making Europe a pole for excellence on corporate social responsibility*, Com(2006)136, final, Bruxelles.

EU. (2010). *European alliance for families, promoting family-friendliness at workplace through awards, audits and labelling schemes*. Retrieved at http://ec.europa.eu/employment_social/emplweb/families/admintool/userfiles/file/Bosnicovareport-formatted.pdf.

European Alliance for CSR. (2008). *Valuing non-financial performance: A European framework for company and investor dialogue*. Author.

Flynn, G. (2008). The virtuous manager: A vision for leadership in business. *Journal of Business Ethics, 78*(3), 359–372.

Gartner, W. B. (2007). Entrepreneurial narrative and a science of the imagination. *Journal of Business Venturing, 22*, 613–627.

Globe. (2008). *Global leadership and organizational behavior effectiveness*. Culture and leader effectiveness. Retrieved July 23, 2013, from www.ccl.org/leadership/pdf/GlobeStudy.

GPTW. (2013). *Great place to work Institute Italia. Best workplaces italiani nel 2014. Magazine di approfondimento sulle migliori aziende per cui lavorare in Italia nel 2014*. Milano: GPTW Inc. Available at: http://www.greatplacetowork.com; energy.loccioni.com/2013/12/we-are-a-great-place-to-work/; www.loccioni.com/2013/.../loccioni-unica-impresa-italiana-sul-podio-del-great-place-

Gray, R., Owen, D., & Adams, C. (1996). *Accounting and accountability: Changes and challenges in corporate social and environmental reporting*. London: Prentice-Hall.

Gray, R. H., Adams, C., & Owen, D. (2014). *Accountability, social responsibility and sustainability. Accounting for society and the environment*. Pearson: Harlow.

Greenberg, J., & Colquitt, J. A. (Eds.). (2005). *Handbook of organizational justice*. Mahwah, NJ: Lawrence Erlbaum.

Gröschl, S. (Ed.). (2013). *Uncertainty, diversity and the common good. Changing norms and new leadership paradigms* (pp. 25–40). Surrey, England: Gower.

Hancock, K. (2005). Employee engagement partnerships: Can they contribute to the development of an integrated CSR culture? In CPI (Cambridge Programme for Industry) 2005. Partnership Matters: Current issues in cross-sector collaboration. Issue 3: 7–20. Retrieved at www.cpi.cam.ac.uk/pccp/ParternMatters3.pdf.

Hartman, E. M. (1998). The role of character in business ethics. *Business Ethics Quarterly, 8*(3), 547–559.

Hartman, E. (2011). Virtue, profit and the separation thesis: An Aristotelian view. *Journal of Business Ethics, 99*, 5–17.

Hoivik von Weltzien, H. (2014). The heart of leadership is ethics. *Impresa Progetto-Electronic Journal of Management, 11*, 1–9.

Hoivik von Weltzien, H., & Melé, D. (2009). Can a SME become a global corporate citizen? Evidence from a case study. *Journal of Business Ethics, 88*, 551–562.

IFERA. (2003). Family firms dominate. *Family Business Review, 16*(4), 235–239.
Istao. (2014). Marchegian excellent companies—Classifica delle principali imprese marchigiane, anno 2013. Ancona: Istao, Fondazione Aristide Merloni, Ottobre 2014. Retrieved September 20, 2014, from http://www.fondazione-merloni.it.
Jacobsen, C. (2001). Dynamics of charismatic leadership: A process theory, simulation model, and tests. *Leadership Quarterly, 12*(1), 75.
Ketola, T. (2008). A holistic corporate responsibility model: Integrating values, discourses and actions. *Journal of Business Ethics, 80*, 419–435.
Koos, S. (2012). The institutional embeddedness of social responsibility: A multilevel analysis of smaller firms' civic engagement in Western Europe. *Socioeconomic Review, 10*(1), 135–162.
Liu, C. H. (2007). *Transactional, transformational, transcendental leadership: Motivation effectiveness and measurement of transcendental leadership.* Paper presented at leading the future of the public sector: The third transatlantic dialogue University of Delaware, Newark, Delaware, USA, May 31–June 2, 2007.
Loccioni Group. (2008a). *La misura di un sogno: L'avventura di Camillo Olivetti—The measure of a dream: The adventure of Camillo Olivetti. Cultura d'Impresa* (Vol. 2). Osimo: Grafiche Scarponi.
Loccioni Group. (2008b). *Un discorso di Adriano Olivetti ai dipendenti. Cultura d'Impresa* (Vol. 2). Osimo: Grafiche Scarponi.
Loccioni Group. (2008c). *Società della conoscenza e attualità del modello Olivetti. Cultura d'Impresa* (Vol. 4). Osimo: Grafiche Scarponi.
MacIntyre, A. (1985). *After virtue. A study in moral theory* (2nd ed.). London: Duckworth.
Malloch, T. R. (2009). *Spiritual enterprises. Doing virtuous business.* New York: Encounter Books.
Martin R. L. (2002). The virtue matrix: Calculating the return on corporate responsibility. *Harvard Business Review*, March, 5–11.
McElroy, J. C. (2001). Managing workplace commitment by putting people first. *Human Resource Management Review, 11*(3), 327–336.
Mees, A., & Bonham, J. (2004). Corporate social responsibility belongs with HR. *Canadian HR Reporter,* April 5, 11–13.
Melcrum. (2006). *Engaging employees in corporate responsibility: How the world's leading companies embed CR in employee decision-making.* London: Author.
Melé, D. (2002). *Not only stakeholder interests. The firm oriented toward the common good.* Notre Dame: Notre Dame University Press.
Melé, D. (2006). Religious foundations of business ethics. In M. J. Epstein & K. O. Hanson (Eds.), *The accountable corporation* (pp. 11–43). London: Praeger Publishers.
Melé, D. (2009a). *Business ethics in action. Seeking human excellence in organizations.* Hampshire: Palgrave Macmillan.
Melé, D. (2009b). Integrating personalism into virtue-based business ethics: The personalist and the common good principles. *Journal of Business Ethics, 88*, 227–244.
Melé, D. (2012). The firm as a "community of persons": A pillar of humanistic business ethos. *Journal of Business Ethics, 106*(1), 89–101.
Novak, M. (1996). *Business as a calling: Work and the examined life.* New York: Simon and Schuster, Free Press.
Nussbaum, M. (2000). *Women and human development: The capabilities approach.* Cambridge: Cambridge University Press.
Olsen, O. K. (2010). *Are good leaders moral leaders? The relationship between effective military operational leadership and morals* Dissertation for the philosophiae doctor degree (PhD) at the University of Bergen.
Opdebeeck, H. (2013). The role of the charismatic economist E.F. Schumacher in economic and civil life: CSR and beyond. In L. Bruni & B. Sena (Eds.), *The charismatic principle in social life* (pp. 151–170). New York: Routledge.
Oreg, S., & Berson, Y. (2011). Leadership and employees' reactions to change: The role of leaders' personal attributes and transformational leadership style. *Personnel Psychology, 64*(3), 627–659.

Painter-Morland, M. (2008). Systemic leadership and the emergence of ethical responsiveness. *Journal of Business Ethics, 82*, 509–524.
Palshikar, K. (2007). Charismatic leadership. Retrieved from http://www.unc.edu/~ketan/documents/Charismatic%20Leadership.pdf.
Pieper, J. (1966). *The four cardinal virtues: Prudence, justice, fortitude, temperance*. Notre Dame, IN: University of Notre Dame Press.
Pirson, M., & Turnbull, S. (2011). Toward a more humanistic governance model: Network governance structure. *Journal of Business Ethics, 99*(1), 101–114.
Rest, J., & Narvaez, D. (1994). *Moral development in the professions. Psychology and applied ethics*. Hillsdale, NJ: Lawrence Erlbaum.
Röpke, W. (1960). *A humane economy*. Chicago: Henry Regnery.
Ruisi, M. (2010). *Measure entrepreneurial virtues. Towards a new perspective for the indicators of corporate success*. Paper presented at the 23rd Eben annual conference *which values for which organisations*?, Trento, 9–11 September.
Ruisi, M., Fasone, V., & Paternostro, S. (2009, April 17–18). *Respect and hope a binomial relationship supporting the creation of a true entrepreneurial model*. Paper presented at *Respect and economic democracy*, Fifth annual conference of the European SPES Forum, Catania.
Schein, E. H. (1983). The role of the founder in creating organizational culture. *Organizational Dynamics*, Summer, 13–28.
Sen, A. (1999). *Development as freedom*. New York: Random House.
Simons, T. L. (1999). Behavioral integrity as a critical ingredient for transformational leadership. *Journal of Organizational Change Management, 12*, 89–104.
Solomon, C. R. (1992a). *Ethics and excellence: Cooperation and integrity in business*. New York: Oxford University Press.
Solomon, R. C. (1992b). Corporate roles, personal virtues: An Aristotelian approach to business ethics. *Business Ethics Quarterly, 2*(3), 317–339.
Sorci, C. (Ed.). (2007). *Lo sviluppo integrale delle aziende*. Milano: Giuffré.
Treviño, L. K., & Brown, M. E. (2004). Managing to be ethical: Debunking five business ethics myths. *Academy of Management Executive, 18*(2), 69–81.
Treviño, L. K., Butterfield, K. D., & McCabe, D. (1998). The ethical context in organizations: Influences on employee attitudes and behaviours. *Business Ethics Quarterly, 8*(3), 447–476.
Unioncamere. (2010). *Le Piccole e Medie imprese nell'economia italiana. Rapporto 2009*. Istituto Tagliacarne. Milano: F.Angeli.
Valentine, S., & Fleishman, G. (2008). Ethics programs, perceived corporate social responsibility and job satisfaction. *Journal of Business Ethics, 77*(3), 159–172.
Varvelli, M. R., & Varvelli, R. (2014). *2Km di futuro. L'impresa di seminare bellezza*. Milano: Il Sole 24 Ore.
Visser, W. (2011). The nature of CSR leadership. Definitions, characteristics and paradoxes. *CSR International Paper Series, 4*, 1–10.
Von Ahsen, A. (2015). Sustainability leadership. In S. O. Idowu, N. Capaldi, M. Fifka, L. Zu, & R. Schmidpeter (Eds.), *Dictionary of corporate social responsibility* (p. 444). Heidelberg: Springer.
Walker, T., & Schmidpeter, R. (2015). Maturity model of CSR. In S. O. Idowu, N. Capaldi, M. Fifka, L. Zu, & R. Schimdpeter (Eds.), *Dictionary of corporate social responsibility* (pp. 312–313). Heidelberg: Springer.
Weaver, G. (2004). Ethics and employees: Making the connection. *Academy of Management Executive, 18*(2), 121–125.
Weaver, G., Trevino, L. K., & Cochran, P. L. (1999). Integrated and decoupled corporate social performance: Management commitments, external pressures and corporate ethics practices. *Academy of Management Journal, 42*(5), 539–552.
Webley, S., & LeJeune, M. (2005). *Corporate use of codes of ethics*. London: Institute of Business Ethics.
Weidinger, C., Fischler, F., & Schmidpeter, R. (2013). *Sustainable entrepreneurship*. Heidelberg: Springer.

Werhane, P. (1999). *Moral imagination and management decision making*. Oxford: University Press.

Whitener, E. M. (2001). Do "high-commitment" human resource practices affect employee commitment? A cross-level analysis using hierarchical linear modeling. *Journal of Management, 27* (5), 515–535.

Yin, R. K. (1994). *Case study research: Design and methods* (2nd ed.). Thousand Oaks, CA: Sage.

Zamagni, S. (Ed.). (1995). *The economics of altruism*. Hants: E. Elgar.

Zimbardo, P. G. (2007). *The Lucifer effect: Understanding how good people turn evil*. New York: Random House.

Zsolnai, L. (Ed.). (2002). *Ethics in the economy. Handbook of business ethics*. Oxford: Peter Lang Academic.

Part IV
CSR and Asia

Does Foreign Ownership Enhance the Corporate Social Performance of Japanese Firms?

Megumi Suto and Hitoshi Takehara

1 Introduction

Given the globalisation of business and widespread changes in organisational ownership structures in the last decade, information related to the social and environmental elements of corporate activities has become indispensable for valuating a corporation. Related to this, it has become increasingly important for corporate managers of global businesses to understand the association between corporate financial performance (CFP) and corporate social responsibility (CSR) and the ways in which this relationship shapes corporate social performance (CSP) in a corporate governance framework.

Numerous studies in the domain of strategic corporate management have argued for an alignment of business objectives with CSR by facilitating the firm's adaptation to a constantly changing society and corporate environment. Stakeholder theory provides a useful theoretical framework of corporate governance for linking businesses' responsibilities with corporate value creation. Conflicts between corporate stakeholders' objectives impede corporate value creation and exacerbate information asymmetry among shareholders.

This study represents an attempt to utilize stakeholder theory to explore the relationship between ownership type and CSP in the Japanese context. More specifically, the purpose of this study is to examine the influence of foreign ownership (relative to domestic ownership) on CSP and the attributes of Japanese firms in the late 2000s. The globalisation of business ownership structures has given foreign investors a dynamic role in transforming the insider-oriented corporate governance

M. Suto (✉) • H. Takehara
Graduate School of Business and Finance, Waseda University, 1-6-1 Nishiwaseda, Shinjuku-ku, Tokyo 169-8050, Japan
e-mail: megumi.suto@waseda.jp; takehara@waseda.jp

H. Lu et al. (eds.), *Building New Bridges Between Business and Society*, CSR, Sustainability, Ethics & Governance, DOI 10.1007/978-3-319-63561-3_11

structure to a more open and transparent structure by motivating firms alter the ways in which they manage stakeholders and relate with shareholders.

This paper is organised as follows. In Sect. 2, offering a short literature review related to the relationships between ownership structure and CSR, we discuss potential CSR problem for Japanese firms in globalization of ownerships and present this study's research hypotheses. In Sect. 3, we explain our methods for constructing the CSP indices and other data and selecting our sample. In Sect. 4, we show and discuss the regression results in which we examine the relationship between ownership structure and CSP. Section 5 concludes and provides some policy implications for the development of CSR practices.

2 Background and Hypotheses

2.1 Stock Ownership Structure and Corporate Social Performance

Within the domain of corporate governance, CSR represents a mechanism for aligning corporate management with the interests of stakeholders by promoting sustainability rather than exclusively seeking to enhance shareholder value. Managerial decisions that do not account for stakeholders' concerns may harm a firm's long-term competitiveness, increase business risk, damage employees' motivation, diminish the firm's reputation, and reduce clients' trust in the firm (e.g. Surroca et al. 2010). Failure to incorporate stakeholder input into the development of CSR practices may also amplify financial risks if investors perceive a firm's socially irresponsible behaviour to be an indicator of greater uncertainty related to its future CFP (Ghoul et al. 2011; Goss and Roberts 2011). Empirical research on the association between CFP and CSP using a governance framework provides businesses with practical insight regarding how firms' roles in society take shape in the interactions with their stakeholders (de Graaf and Stoelhorst 2013).

Since 1990s, numerous research has increasingly focused on the influence of institutional ownership on corporate managers' decision-making. Many of them suggests that external shareholders (e.g. institutional investors, foreign investors) can drive corporate managers to improve their firms' CSP and increase the degree to which they engage in related disclosures (Chaganti and Damanpour 1991; Coffey and Fryxell 1991; Graves and Waddock 1994; Johnson and Greening 1999; Cox et al. 2004; Neubaum and Zahra 2006). Generally, these studies have tended to find that in the US and the UK markets there exists a positive relationship between institutional shareholdings and CSP.

Corporate social responsibility (CSR) is a core topic within the domain of corporate governance of global businesses (Brickley et al. 2003; Deakin and Hobbs 2007; de Graaf and Stoelhorst 2013). It has become important for corporate governance studies to explore the effects of changing ownership structures on corporate social performance to address issues related to the development of global businesses and ultimately increase cross-border investment (Cox and Schneider 2010; Dam and Scholtens 2012, 2013). As cross-border portfolio investment has continued to grow, institutional investors have generally showed a preference for firms with low transaction costs or high liquidity (Gompers and Metrick 2001) and have generally chosen large, esteemed firms rather than firms that are small or poorly governed as targets for investment (Leuz et al. 2009). Moreover, institutional investors demonstrate a home-country bias in selecting targets for investment (Choe et al. 2005; Leuz et al. 2009). Many studies have found that US investors show a strong preference for disclosure and transparency. As such, they tend to avoid insider-trading systems, peculiar relationships, and weak stockholder protection (Aggarwal et al. 2005; Kang and Stulz 1997; Kho et al. 2006; Leuz et al. 2009). Therefore, foreign investors tend to invest in firms with good corporate governance, high social trust, and a good reputation to circumvent problems associated with information asymmetry.

2.2 *Stock Ownership Structure in Japan*

Traditional Japanese corporations are characterised by insider-oriented system with the unique feature of corporate ownership structure named cross-shareholding based on long-term business relationships and internal governance mechanism based on long-term employment and seniority system. Their self-discipline or CSR policy is typically passed down in the business over generations, focusing on the assurance of product quality, the contribution to social causes, the provision of employment to the community. These relationship-based shareholdings may mitigate information asymmetry among firms and financial institutions, but increase information asymmetry with outside shareholders.

The cross-shareholding has declined since the late 1980s when the bubble economy burst and financial business slumped. Moreover, foreign investors have emerged as major shareholders since the 1990s, when globalisation of business and financial liberalisation proliferated in conjunction with worldwide growth of capital flow and cross-border diversification of portfolio investments (Ahmadjian 2007). In parallel with a decline in cross-shareholdings, the performance of long-term institutional investors (e.g. pension funds) became a growing topic of interest in an aging society.

According to Tokyo Stock Exchange Stock Ownership Survey on all listed companies in Japan, in 1990, foreign investors accounted for 4.7% of shareholding of Japanese firms. This figure rose to 18.8% in 2000 and peaked at 28% in 2006. After a temporary drop resulting from the global financial crisis in 2008, the level recovered in 2012. Foreign investors become a major player in the market. In contrast, shareholding by business corporations and financial intermediaries fell from 41.8% in 2000 to 31.2% in 2012. Domestic institutional investors (e.g. pension funds) kept about 19% through the 2000s have remained silent until recently. Shareholding by individuals held steady at about 20% since the end of 1980s.

In the situation, foreign investors have faced with substantial information costs in the market and weak corporate governance among potential investee firms (Miyajima and Nitta 2011). Despite gradual institutional extension of non-financial disclosure[1], the transparency of Japanese firms in the late 2000s remained insufficient from a global perspective (e.g. Stewart and Yermo 2010; Asian Corporate Governance Association 2008). In globalisation of business and ownership structures, relationship-based shareholdings increase information asymmetry with outside shareholders. Nevertheless, disclosure of non-financial information and accountability for multi-facet aspects of their business remain insufficient, which may weaken their conventional self-governance mechanisms and degrade trust in their business.

Given this, we seek to address whether changing patterns of ownership in the late 2000s motivated firms to change their corporate governance strategies toward less insider-oriented and reconstruct trust of foreign investors.

2.3 Hypotheses Development

Foreign corporate investors, who possess less information than their domestic counterparts, may show a greater preference for CSP corporations willing to consider the non-financial elements of investees' businesses to avoid risk or to reduce agency costs. In relationship-based system, domestic corporate investors may consider CSR practices to be intrinsically tied to higher costs or have little bearing on financial performance. Further, in the Japanese market, long-term institutional investors such a pension funds were unconcerned with CSP and tended to be subject to a short-term bias in the stagnant economy in the early 2000s (Suto and Toshino 2005). Foreign investors that engage in value-enhancing or value-seeking strategies may demonstrate a stronger preference for firms that perform well in terms of their social responsibilities than domestic counterparts.

We propose the following hypotheses to explore the relationship between different types of ownership and the preference to invest in Japanese firms with high CSP in the late 2000s. It is important for corporate managers to understand what dimensions of CSR a type of investors prefer, as CSR is multi-facet. Given this, we develop the following hypotheses on both each dimensional CSPs and Composite CSP.

Hypothesis 1 *There is a* positive *relationship between foreign ownership and CSP.*

Hypothesis 2 *The positive relationship between foreign ownership and CSP is more pronounced than the relationship between domestic ownership and CSP.*

---

[1]In 2006, the Tokyo Stock Exchange required listed companies to disclose a Corporate Governance Report. In 2008, the Financial Instruments and Exchange Act required corporations to submit Internal Control Reports to the Ministry of Finance.

The fundamental purpose of this paper is to examine whether increases in foreign ownership enhances firms' CSP. To identify the respective influences of foreign investors on CSP both dimensionally and comprehensively (relative to domestic investor groups) in the Japanese stock market, we offer the following hypotheses. These hypotheses represent the central focus of this study:

Hypothesis 3 *Increases in foreign ownership enhance CSP.*

Hypothesis 4 *Increases in domestic ownership less enhance CSP than foreign ownership.*

3 Data and Sample

3.1 Construction of CSP Indices

Because the management of stakeholders' competing interests can contribute to successful economic performance, CSP can be effectively analysed by adopting a stakeholder management perspective (Barnea and Rubin 2010; Berman et al. 1999; Clarkson 1995; Harrison and Freeman 1999; Scholtens and Zhou 2008). In this study, we first define CSP dimensions from a stakeholder management perspective and define Composite CSP to integrate these dimensions. We identify five CSP dimensions according to stakeholder relations: employee relations (EMP), social contributions (SC), organisation security and product safety (SS), internal governance and risk management (IG), and environmental preservations (ENV).

Effective employee relations can improve the quality of work produced by employees and employee motivation. Social contributions relate to a firm's policy for, and response to, social demands. Firm security and product safety is related to customer confidence in the quality of a firm's products and trust in the transactions in which it engages. Internal governance and risk management concerns the firm's disclosure policy for investors, including organisational information sharing, compliance and self-disciplining. Environmental preservation represents a pillar of CSR in a society that has grown increasingly concerned with global climate change.[2]

We constructed the CSP indices related to these five dimensions with a principal component analysis using the Toyo Keizai Corporate Social Responsibility Database as a primary data for the period 2007–2011.[3] We computed

[2]This classification of stakeholders is consistent with the Girerd-Potin et al. (2014) typology: business stakeholders (employee, customers, and suppliers), social stakeholders (the environment and society), and financial stakeholders (stockholders and debt holders) and is also consistent with Clarkson's (1995) classification of primary stakeholders.

[3]Toyo Keizai Incorporated have annually sent the questionnaire to the listed firms in the beginning of July and retrieved responses by the end of September since 2006. However, the form of the questionnaire was significantly revised in 2007. The original database consists of three parts: employee relations (Part I), an overall survey related CSR (Part II), and environmental

the composite CSP index on the basis of the five dimensional indices. Let $r(\ )$ denote the function that gives a rank of the element of the vector in ascending order and n denote the number of firms in each year. Given this, the comprehensive measure of CSP in each year is defined as follows. Because our CSP dimensional indices approximately obey a standard normal distribution, we adjusted Eq. (1) such that the composite measure of CSP is uniformly distributed and falls in the closed interval [-3, 3].

$$CSP = \frac{r(r(EMP) + r(SC) + r(SS) + r(IG) + r(ENV)) - 1}{n - 1} \times 6 - 3 \quad (1)$$

3.2 Categorization of Ownership and Control Variables

We categorised investors in the Japanese stock market into three groups of owners within the constraint of available data: domestic corporate investors, foreign corporate investors, and domestic individual investors. For the purposes of this study, shareholding by domestic corporations is defined as the sum of shares owned by financial institutions and shares owned by other corporations, excluding shares owned by securities brokers, governments, and public organisations.[4]

To control for firm characteristics that may inadvertently affect the relationship between stock ownerships and CSP, we employed eight control variables; firm size, profitability, credit risk, and growth rate, turnover rate in the stock market, Book-to-Market ratio, and foreign dependency ratio. The primary source

preservation (Part III). We subdivided Part II into three distinct CSR dimensions, which correspond to the stakeholder relations we chose. First, we selected 17 questions related to employee relations, 21 questions concerning CSR in a general sense, and 18 questions regarding environmental preservation. We first converted quantitative data (e.g. proportion of female employees) to three- or four-level categorical data. Then, we made within-sector adjustments because some questions had different meanings among sectors. For each of the five CSP attributes, we used a principal component analysis to construct CSP dimensional indices. On the basis of responses to the questions, we kept 13 scores regarding employee relations (EMP), five scores regarding social contributions (SC), five scores regarding security of the firm and product safeness (SS), six scores regarding internal governance and risk management (IG), and five scores regarding environmental preservation (ENV). The item scores and their related factor loading are shown in Appendix 1. We then demeaned and scaled each CSP dimensional index by its standard deviation so that it approximately obeyed a standard normal distribution.

[4]Japanese corporations are required to disclose a summary of their stock ownership structure in their financial reports. This summary describes the number of shares owned by domestic corporations, foreign corporations, and individual investors. This categorisation scheme is restricted by the amount of information that firms publish, but it is sufficient for analysing foreign ownership preferences for and influences on CSP relative to domestic ownership. Thus, this study represents a viable first step in investigating the link between ownership structure and CSR activities in Japanese firms.

for financial data is the NIKKEI NEEDS Database and the Financial Data Solutions NPM Database.

We used the natural logarithm of a firm's total assets (in million JPY), lnTA, as a measure of firm size. We also included proxies for profitability, credit risk, and growth of the firm as control variables. These variables were Return on Assets (ROA), Debt Ratio (DR = total debt/total asset), and Growth Rate of Total Assets (GTA). In addition, to account for a firm's liquidity and variability of stock price (to which institutional investors attach importance), we also included variables related to monthly turnover rate (Turn) and 36-month historical volatility (Vol3Y), respectively and the Book-to-Market ratio (B/M) in the model to control for differences in portfolio style (i.e. value vs. growth). Finally, we incorporated the Foreign Dependency Ratio (FDR) variable, which was defined as sales in foreign countries divided by total sales.

3.3 Sample Selection and Preliminary Analyses

Our selection of sample firms was contingent upon the availability of relevant data including data regarding CSP, ownership, and financial statement for the period between 2007 and 2011. All firms included in our sample were non-financial listed firms on the stock exchanges during this period[5] (Table 1). The number of firms ranged from 753 in 2007 to 868 in 2011. About 60% of the firms are listed on the TSE first section; however, roughly 10% of the sample firms are listed on the

Table 1 Number of sample firms

Sector	2007	2008	2009	2010	2011	TSE1	TSE2	Others	Total
Consumption goods	202	192	197	213	215	206	29	47	282
Investment goods	286	297	309	312	310	298	41	72	409
Services	209	225	254	273	279	207	40	148	388
Transportation	19	20	20	21	25	22	3	4	29
Utility	11	12	13	15	11	15	0	0	15
Real estate	26	29	23	27	28	28	6	12	45
All sectors	753	775	816	861	868	776	119	283	1168

Number of firms sampled at the end of September of each year (2007–2011) and number of firms listed on the Tokyo Stock Exchange 1st Section (TSE1), on Tokyo Stock Exchange Second Section (TSE2), and other stock exchanges in Japan (Others). Number of firms in the most right four columns is non-duplicated and a single firm appears four times at maximum in our sample period, 2007 through 2011

[5]The format of financial statement of financial firm differs substantially from that of non-financial firms. Therefore, the economic meanings of items in financial statement such as net income and assets are difference between non-financial firms and financial firms. Because return and risk measures we employed in this study are not comparable between two types of firms, we excluded financial firms in the empirical analysis.

second section of the TSE and 30% of sample firms are listed on exchanges other than the TSE.

Preliminarily, we calculated correlations between the CSP indices and the ownership structure as well as the characteristics variables with the pooling data (Table 2). Panel A of Table 2 shows the Spearman rank correlations between CSP and ownership structure; Panel B shows the correlations between CSP and the eight firm characteristic variables outlined above. As indicators of ownership structure, we employed both current levels and past 5-year-change in the share ownership.

In Panel A, the correlations between the CSP indices and shares owned by both foreign and domestic corporate investors are positive and significant at 1% level. The correlation between the CSP indices and 5-year increase in foreign ownership is also significant and positive, but the analogous correlation related to domestic corporate ownership is significant and negative. There exists significant variation in the magnitudes of the correlations between the CSP dimensional indices and stock ownership. Individual ownership is positively related to neither the composite CSP nor the CSP dimensions. These findings imply that foreign investors may prefer firms with superior CSP than domestic investors and suggest that they may have increased investment in high CSP firms.

Table 2 Correlation among CSP, ownership structure and firms' characteristics

Panel A. Spearman rank correlation between CSP and stock ownership structure

	Foreign investors	Japanese corporations	Individual investors	Past 5 years increase in shares held by foreign investors	Past 5 years increase in shares held by Japanese corporations	Past 5 years increase in shares held by individual investors
Composite CSP	0.535	0.195	−0.485	0.168	−0.120	−0.032
p-value	0.000	0.000	0.000	0.000	0.000	0.040
Employee relations	0.357	0.126	−0.324	0.133	−0.067	−0.062
p-value	0.000	0.000	0.000	0.000	0.000	0.000
Social contribution	0.458	0.179	−0.444	0.136	−0.103	−0.011
p-value	0.000	0.000	0.000	0.000	0.000	0.479
Security and safety	0.335	0.093	−0.270	0.092	−0.070	−0.023
p-value	0.000	0.000	0.000	0.000	0.000	0.138
Internal governance	0.226	0.054	−0.192	0.049	−0.037	−0.001
p-value	0.000	0.001	0.000	0.002	0.019	0.969
Environment	0.496	0.212	−0.466	0.167	−0.152	−0.001
p-value	0.000	0.000	0.000	0.000	0.000	0.967

(continued)

Table 2 (continued)

Panel B. Spearman rank correlation between CSP and firms' characteristics

	lnTA	ROA	DR	GTA	Turn	Vol3Y	BPR	FDR
Composite CSP	0.664	0.080	0.017	0.049	0.409	−0.060	−0.321	0.086
p-value	0.000	0.000	0.276	0.002	0.000	0.000	0.000	0.000
Employee relations	0.438	0.066	0.030	0.033	0.264	−0.035	−0.230	0.074
p-value	0.000	0.000	0.056	0.033	0.000	0.026	0.000	0.000
Social contribution	0.604	0.058	0.039	0.061	0.337	−0.067	−0.276	0.047
p-value	0.000	0.000	0.014	0.000	0.000	0.000	0.000	0.003
Security and safety	0.361	0.068	−0.050	0.023	0.255	−0.057	−0.216	0.103
p-value	0.000	0.000	0.001	0.149	0.000	0.000	0.000	0.000
Internal governance	0.271	0.036	−0.002	0.015	0.192	0.001	−0.150	−0.017
p-value	0.000	0.023	0.881	0.325	0.000	0.933	0.000	0.281
Environment	0.652	0.043	0.048	0.044	0.378	−0.057	−0.255	0.105
p-value	0.000	0.006	0.002	0.005	0.000	0.000	0.000	0.000

lnTA natural logarithm of total asset (in million JPY), *ROA* Return of Asset, *DR* Debt ratio, *GTA* Growth rate in total asset, *Turn* Monthly turnover, *Vol3Y* Past 3 year volatility of monthly stock returns, *BPR* Book-to-price ratio, *FDR* Foreign dependency ratio defined as (sales in foreign countries)/(total sales)

Panel B confirms the relationships between firm characteristics and the CSP indices. The correlations between composite CSP and the characteristics variables (with the exception of DR) are statistically significant at 1% level. As a whole, the results demonstrate that high CSP firms are typically large-scale and operate globalised business, with high liquidity and high performance.

4 Regression Analyses

4.1 Relationship Between Stock Ownership and CSP

To examine the relationship between stock ownership and CSP, we conducted a multivariate regression analysis in which we employed the control variables discussed above.

$$y_{j,t} = \alpha + \beta x_{j,t} + \gamma_1 ROA_{j,t} + \gamma_2 DR_{j,t} + \gamma_3 GTA_{j,t} + \gamma_4 Turn_{j,t} + \gamma_5 Vol3Y + \gamma_6 BPR_{j,t} + \gamma_7 FDR_{j,t} + \sum_{i=2}^{3} \delta_i DSize_{i,j,t} + \sum_{i=2}^{6} \lambda_i DSecter_{i,j} + \sum_{t=2007}^{2010} \eta_t DYear_{j,t} + \varepsilon_{j,t}. \quad (2)$$

Dependent variable $y_{j,t}$ is a composite CSP or one of the five CSP dimensional indices of firm *j* in year *t*. Independent variable $x_{j,t}$ is one of shares owned by foreign corporations, shares owned by Japanese corporations, and shares owned by individuals, of firm *j* in year *t*. ROA, DR, GTA, Turn, Vol3Y, BPR, and FDR are control variables. Since we observed non-linear relationship between CSP and firm size, we also constructed three size dummy variables—Size1, Size2, and Size3—to incorporate into the regression analysis. $DSize_{i,j,t}$ is a size dummy which is equal to 1 if firm *j* belongs to the *i*-th size ranked portfolio in year *t*, and 0 otherwise. $DSector_{i,j}$ is a sector dummy variable which is equal to 1 if firm *j* belongs to the *i*-th sector, and 0 otherwise. Finally, $DYear_{j,t}$ represents a dummy variable to indicate each year we evaluated, where $t = 2007, \ldots, 2010$.

To mitigate endogeneity resulting from potential possible reverse causality between CSP and ownership variables, we performed two-stage least square regression analysis in which we used a 1-year lagged ownership variable and a dummy (NOTSE1) as instrument variables.[6] NOTSE1 adopts the value of 1 if the firm is not listed in the TSE First Section, and 0 if the firm is listed in the First Section.[7]

Table 3 reports the results of the regression analyses for different investor groups with both Composite CSP and five CSP dimensional indices serving as the dependent variables. Foreign ownership has significantly positive relationships with Composite CSP and all CSP dimensions and the coefficients are higher than those for domestic corporate ownership and individual ownership as well. These results support Hypothesis 1 and 2.

[6]We first ran an OLS analysis and considered observations whose standardised residuals were larger than 3.0 or smaller than −3.0 to be outliers. In the subsequent two-stage LS analysis, we excluded these observations. When we computed the *t*-values for regression slopes, standard errors were corrected by the two-way cluster error correction method described by Petersen (2009).

[7]Since we are using short panel data in this research, we employ a regression model with sector dummy and year dummy variables instead of two way fixed effects model to avoid a large decrease in the degree of freedom. We conducted the Wu-Hausman's test for endogeneity and Sargan's over-identification test before the two-stage least-square analysis. The results of these tests are available upon request from the authors. Since Wu-Hausman's test statistics are not significant at 5% level in most cases, endogeneity is not severe in regression models (2) and (3), though we use a two-stage regression method.

Table 3 Effects of stock ownership structure on the CSP

	CSP	EMP	SC	SS	IG	ENV
Panel A. % Shares held by foreign corporations						
Intercept	1.929***	0.844***	0.867***	0.951***	0.376***	0.777***
Foreign Corp.	0.014***	0.006***	0.011***	0.004***	0.005***	0.006***
ROA	−0.001	0.005	−0.004	0.001	−0.006*	0.000
DR	−0.002**	0.000	0.000	−0.001*	−0.002**	0.000
GTA	−0.005*	−0.003	0.000	−0.001	−0.001	−0.003*
Turn	0.002**	0.002***	−0.001	0.002***	0.000	0.001*
Vol3Y	−0.034***	−0.032***	−0.012***	−0.028***	0.007*	−0.016***
BPR	−0.002***	−0.001***	−0.001***	−0.001***	−0.001***	0.000
FDR	0.003**	0.001	0.001	0.002**	−0.001	0.001*
Adjusted R^2	0.457	0.200	0.389	0.206	0.088	0.397
Panel B. % Shares held by domestic corporations						
Intercept	1.787***	0.740***	0.901***	0.994***	0.499***	0.668***
Domestic Corp.	0.008***	0.004***	0.005***	0.001	0.001	0.005***
ROA	0.000	0.007	−0.005	−0.001	−0.006*	0.000
DR	−0.005***	0.000	−0.002***	−0.002***	−0.002***	−0.001**
GTA	−0.006**	−0.005**	0.000	−0.002	0.000	−0.003**
Turn	0.005***	0.003***	0.001	0.003***	0.000	0.002***
Vol3Y	−0.037***	−0.032***	−0.014***	−0.029***	0.007*	−0.016***
BPR	−0.002***	−0.001***	−0.001***	−0.001***	−0.001***	0.000
FDR	0.005***	0.003**	0.002*	0.003***	0.000	0.002***
Adjusted R^2	0.440	0.190	0.376	0.190	0.082	0.399

(continued)

Table 3 (continued)

	CSP	EMP	SC	SS	IG	ENV
Panel C. % Shares held by individual investors						
Intercept	2.522***	1.123***	1.335***	1.104***	0.558***	1.067***
Individuals	−0.012***	−0.005***	−0.008***	−0.002**	−0.002***	−0.007***
ROA	−0.001	0.004	−0.005	0.002	−0.007**	−0.001
DR	−0.004***	0.000	−0.001*	−0.002***	−0.002***	−0.001
GTA	−0.006**	−0.004*	0.000	−0.002	0.000	−0.003**
Turn	0.004***	0.003***	0.000	0.003***	0.000	0.002***
Vol3Y	−0.034***	−0.033***	−0.012***	−0.026***	0.006	−0.016***
BPR	−0.002***	−0.001***	−0.001***	−0.001***	−0.001***	0.000
FDR	0.004***	0.002	0.002*	0.002***	−0.001	0.002**
Adjusted R^2	0.468	0.202	0.404	0.216	0.089	0.408

[Dependent Variables] *CSP* Total CSP, *EMP* Employee relations, *SC* Social Contribution, *SS* Security of the firm and Safeness of the product, *IG* Internal Governance and Risk Management, *ENV* Environment preservations

[Independent Variables] *lnTA* natural logarithm of total asset (in million JPY), *ROA* Return of Asset, *DR* Debt ratio, *GTA* Growth rate in total asset, *Turn*: Monthly turnover, *Vol3Y*: Past 3 year volatility of monthly stock returns, *BPR* Book-to-price ratio, *FDR* Foreign dependency ratio defined as (sales in foreign countries)/(total sales)

*** Significant at $p < .01$, ** Significant at $p < .05$, *Significant at $p < .10$

4.2 Enhancement of Corporate Social Performance by Foreign Investors

Next, we examined the relationship between changes in ownership and changes in CSP scores. It is important to consider that improvements in CSP may take several years, despite managers' immediate reactions to changes in ownership structure. Because our CSP dimensional indices only cover 5 years (2007–2011), it is difficult to explore the causal relationship between long-run changes in firm ownership structure and tangible changes to CSR-related activities. Considering the constraints on the analysis derived from the short-term nature of our model, we employ another regression model.

$$\begin{aligned}\Delta CSP_{j,2008:2011} = \; & \alpha + \beta \Delta OWN_{j,(t-3):t} + \gamma_1 ROA_{j,t} + \gamma_2 DR_{j,t} \\ & + \gamma_3 GTA_{j,t} + \gamma_4 Turn_{j,t} + \gamma_5 Vol3Y + \gamma_6 BPR_{j,t} \\ & + \gamma_7 FDR_{j,t} + \sum_{i=2}^{3} \delta_i DSize_{i,j,t} + \sum_{i=2}^{6} \lambda_i DSecter_{i,j} \\ & + \sum_{t=2007}^{2010} \eta_t DYear_{j,t} + \varepsilon_{j,t}.\end{aligned} \quad (3)$$

The basic structure is illustrated in Fig. 1. First, the dependent variable ΔCSP, is fixed as a 3-year change in CSP in Period (4) from 2008 to 2011. The key explanatory variable, ΔOWN, is a 3-year change in stock ownership computed for four distinct time periods. In Period (4), the observation period for ΔOWN coincides with the observation period for ΔCSP. We explored how a changing ownership structure over the past 3 years, Period (1), (2), and (3) affects future CSP in (4). Therefore, we explicitly assumed that changes in ownership trigger changes in CSP, though part of this effect is exerted immediately. In Period (4), the observation

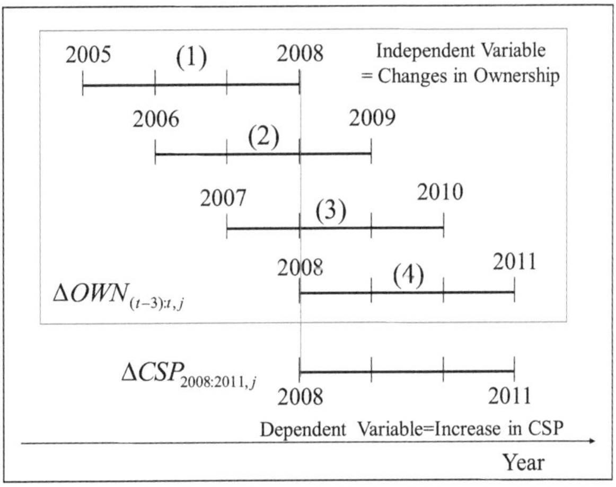

Fig. 1 Regression model to test the relationship between ΔOWN and ΔCSP

period for ΔOWN coincides with the observation period for ΔCSP. Therefore, in this case, we examined the immediate effect of change in ownership on the implementation of CSR-related activities. For these analyses, we incorporated control, instrument, and dummy variables that were identical to those used for model (2).

From regression results in Panel A of Table 4, increases in foreign corporate ownership yielded positive, significant coefficients for the composite CSP variables in Periods (1) and (3) at 10% level. We infer that growing foreign ownership improves the CSP of Japanese firms, and that foreign investors are particularly concerned with the employment relations element of the CSR practices for the firms in which they invest. The coefficients of change in foreign ownership in Period (1) through Period (3) are statistically significant at 5% level. This coincides with our expectations, as we interpret this tendency to indicate that managers require a longer time to improve CSP as it relates to employee relations.

In contrast, Panel B shows increases in domestic corporate ownership, in the most cases, yielded negative, non-significant coefficients for the CSP variables. Among the individual CSR attributes, slopes for EMP are negative and significant in Periods (1) and (2) at 10% level. Panel C shows individual investors may have no significant effect on corporate management except the SC in Period (3) and (4). The relationship between change in individual investor ownership and the SC are negative and significant. It suggests that their distance from corporate management seems substantial relative to corporate investors in general. Taken together, the results support Hypothesis 3 and 4.

Table 4 Impact of changes in shareholdings onto the progress of CSR

Panel A. Case of foreign investors					
3 years increase in shares held by foreign investors					
2008–2011		(1) 2005–2008	(2) 2006–2009	(3) 2007–2010	(4) 2008–2011
⊿CSP	Coef.	0.016**	0.019	0.016*	0.011
	p-value	0.019	0.105	0.085	0.275
⊿EMP	Coef.	0.023***	0.035**	0.024**	0.019
	p-value	0.008	0.014	0.050	0.128
⊿SC	Coef.	0.003	0.008	0.009	0.008
	p-value	0.503	0.178	0.213	0.260
⊿SS	Coef.	0.003	−0.001	0.000	0.001
	p-value	0.352	0.907	0.885	0.738
⊿IG	Coef.	0.003	−0.004	−0.001	0.003
	p-value	0.534	0.637	0.894	0.619
⊿ENV	Coef.	0.002	0.007	0.006	0.003
	p-value	0.733	0.382	0.293	0.661

(continued)

Table 4 (continued)

Panel B. Case of Japanese corporations					
3 years increase in shares held by Japanese corporations					
2008–2011		(1) 2005–2008	(2) 2006–2009	(3) 2007–2010	(4) 2008–2011
⊿CSP	Coef.	−0.005	−0.007	−0.013	0.009
	p-value	0.377	0.415	0.361	0.353
⊿EMP	Coef.	−0.015*	−0.025**	−0.029	−0.002
	p-value	0.079	0.018	0.195	0.851
⊿SC	Coef.	0.003	0.001	0.007	0.013**
	p-value	0.331	0.701	0.312	0.031
⊿SS	Coef.	−0.002	0.003	−0.003	0.001
	p-value	0.437	0.288	0.535	0.710
⊿IG	Coef.	−0.005	−0.001	0.006	0.004
	p-value	0.127	0.775	0.448	0.491
⊿ENV	Coef.	0.003	−0.004	−0.011	−0.008
	p-value	0.426	0.322	0.128	0.133
Panel C. Case of individual investors					
3 years increase in shares held by individual investors					
2008–2011		(1) 2005–2008	(2) 2006–2009	(3) 2007–2010	(4) 2008–2011
⊿CSP	Coef.	−0.005	−0.004	−0.008	−0.013
	p-value	0.391	0.608	0.318	0.113
⊿EMP	Coef.	−0.002	−0.001	−0.003	−0.010
	p-value	0.837	0.914	0.805	0.394
⊿SC	Coef.	−0.004	−0.005	−0.011**	−0.015***
	p-value	0.106	0.137	0.029	0.004
⊿SS	Coef.	0.000	−0.003	0.001	−0.002
	p-value	0.958	0.243	0.742	0.543
⊿IG	Coef.	0.003	0.003	−0.004	−0.005
	p-value	0.357	0.507	0.508	0.352
⊿ENV	Coef.	−0.004	0.000	0.003	0.005
	p-value	0.202	0.931	0.550	0.208

[Dependent variables] ΔCSP is fixed as a 3-year change in CSP in Period (4) from 2008 to 2011
[Explanatory variable] ΔOWN is a 3-year change in stock ownership computed for four distinct time periods
*** Significant at $p < .01$, ** Significant at $p < .05$, *Significant at $p < .10$

5 Conclusions

Of the findings revealed in this study, the most remarkable concerns the evidence produced in relation to the effects of increases in foreign ownership on CSP in Japanese firms. One interpretation of this result is that active behaviour among foreign investors can motivate managers of Japanese firms to review their CSR practices and improve the degree to which they engage in non-financial disclosure in Japanese markets. Even if investment strategies among foreign investors are diverse, emerging concern with CSP by foreign investors can stimulate non-financial disclosures in the Japanese market.

Among the dimensional CSP, we found that foreign investors may be more interested in firms with employment policies that emphasise diversity, work-life balance, and handicapped and aging workers, which have become increasingly important for corporate management in response to the social changes. This finding shows that foreign investors pay particular attention to the internal stakeholder management practices of the firms in which they invest, which is largely consistent with the results of previous studies that showed a positive association between employee relations and firm profitability (e.g. Turban and Greening 1997; Edmans 2011; Faleye and Trahan 2011).

The case of Japanese firms suggests that foreign investors can influence the shift from insider-oriented corporate governance toward a more transparent structure among Japanese firms. The results produced by this study have practical implications for both corporate managers and domestic corporate investors in Japan. Specifically, Japanese firms should be more sensitive to the social and environmental elements of perception of the markets in which they operate. Moreover, firms should improve their disclosure practices in a global business environment. Domestic corporate investors, particularly long-term institutional investors, should have a greater concern with stakeholder management of investee firms and reconsider how they perceive the non-financial aspects of investees' activities in their valuation of corporations.

Despite these findings and implications, this study suffers from some shortcomings associated with the ownership data we used. Our categorisation of investors (i.e. foreign corporate investors, domestic corporate investors, and individuals) fails to distinguish business corporate ownership from institutional ownership or long-term investors from short-term investors. This is notable, given that the aims of these different types of ownership may differ. The further delineation of these ownership structure types will avoid problems associated with heterogeneity. This represents a necessary step for validating the evidence produced by our analyses. By extending this line of inquiry, future researchers can provide a more nuanced understanding of the impact of corporate governance on social responsibility in an increasingly global society.

Appendix. Adopted questions from CSR survey of Toyo Keizai CSR database

	Evaluation point	Weights
Employee relations (EMP)		**29.478**
1	Ratio of female employees to total employees	−0.192
2	Ratio of female managers to total managers	−0.304
3	Ratio of phisically handicapped employees to total employees	−0.282

(continued)

	Evaluation point	Weights
4	Ratio of old employees (60 years old and over) to total employees	−0.252
5	Average years of continuous employment	−0.162
6	Labor turnover rate	−0.349
7	Average salary for a 30 years old	−0.312
8	Overtime hours	−0.328
9	Overtime wage per hour	−0.341
10	Rate of paid holidays taken	−0.344
11	Frequency rates of industrial injuries	−0.223
12	Flexible work arrangement (flex-time, short-time working, on-site child care, etc.)	−0.219
13	Incentive program (internal venture, bonus plan, education program etc.)	−0.208
Social contribution (SC)		**51.736**
1	Comprehensive evaluation (CSR department, director in charge, CSR document etc.)	−0.438
2	Corporate ethics (guidelines, business ethics document, etc.)	−0.263
3	Department of social actions	−0.703
4	Social expenditure per employee	−0.430
5	Matching gift and volunteer grant programs	−0.243
Security of the firm and product safeness (SS)		**45.279**
1	Specialty divisions on investor relations, consumer affairs, cooperation with NPO	−0.268
2	Whistle-blower policy	−0.111
3	Specialty department for managing quality and safety of products and services	−0.910
4	Ratio of domestic business offices with ISO9000 certification	−0.212
5	Ratio of foreign business offices with ISO9000 certification	−0.206
Internal governance and risk management (IG)		**35.766**
1	Comprehensive evaluation (whistle-blower protection, CSR manual, complaint DB, etc.)	−0.151
2	Existence/nonexistence of compliance department	−0.436
3	Existence/nonexistence of CIO	−0.594
4	Existence/nonexistence of CFO	−0.620
5	Information systems (security policy, internal/external auditing etc.)	−0.204
6	Comprehensive evaluation (fair trade, compliance, closedown in the past 3 years, etc.)	−0.093
Environment preservations (ENV)		**49.216**
1	Environmental planning department, director in charge of environmental affairs, etc	−0.496
2	Environmental accounting, disclosure and auditing	−0.587
3	Ratio of environment related business to total revenue	−0.427
4	Promotion of procurement of eco-friendly goods and services	−0.466
5	Ecolabelling (ISO14020 series etc.)	−0.036
6	Environment related compliance (environmental disasters, law violation, etc.)	−0.090

Numbers in the column named ‘Weights’ are the contribution rate (in %) of CSP dimensional indices and loadings of first principal component as of September 2010

References

Aggarwal, R., Klapper, L., & Wysocki, P. (2005). Portfolio preference of foreign institutional investors. *Journal of Banking and Finance, 29*(12), 2919–2946.

Ahmadjian, C. (2007). Foreign investors and corporate governance in Japan. In M. Aoki, G. Jackson, & H. Miyajima (Eds.), *Corporate governance in Japan: Institutional change and organizational diversity* (pp. 125–150). Oxford, UK: Oxford University Press.

Asian Corporate Governance Association. (2008). *White Paper on corporate governance in Japan*. Retrieved December 2012, from http://www.acga-asia.org/public/files/Japan%20WP%20May2008.pdf

Barnea, A., & Rubin, A. (2010). Corporate social responsibility as a conflict between shareholders. *Journal of Business Ethics, 97*(1), 71–86.

Berman, S. L., Wicks, A. C., Kotha, S., & Jones, T. M. (1999). Does stakeholder orientation matter? The relationship between stakeholder management models and firm financial performance. *Academy of Management Journal, 42*(5), 488–506.

Brickley, J. A., Smith, C. W., & Zimmerman, J. L. (2003). Corporate governance, ethics, and organizational architecture. *Journal of Applied Corporate Finance, 15*(3), 34–45.

Chaganti, R., & Damanpour, F. (1991). Institutional ownership, capital structure, and firm performance. *Strategic Management Journal, 12*(7), 479–491.

Choe, H., Koh, B. C., & Stulz, R. M. (2005). Do domestic investors have an edge? The trading experience of foreign investors in Korea. *The Review of Financial Studies, 18*(3), 795–829.

Clarkson, M. B. E. (1995). A stakeholder framework for analyzing and evaluating corporate social performance. *Academy of Management Review, 20*(1), 92–117.

Coffey, B. S., & Fryxell, G. E. (1991). Institutional ownership of stock and dimensions of corporate social performance: An empirical examination. *Journal of Business Ethics, 10*(6), 437–444.

Cox, P., Brammer, S., & Millington, A. (2004). An empirical examination of institutional investor preferences for corporate social performance. *Journal of Business Ethics, 52*(1), 27–43.

Cox, P., & Schneider, M. (2010). Is corporate social performance a criterion in the overseas investment strategy of U.S. pension plans? An empirical examination. *Business and Society, 49*(2), 252–289.

Dam, L., & Scholtens, B. (2012). Does ownership type matter for corporate social responsibility? *Corporate Governance: An International Review, 20*(3), 233–252.

Dam, L., & Scholtens. (2013). Ownership concentration and CSR policy of European Multinational Enterprises. *Journal of Business Ethics, 118*, 117–126. doi:10.1007/s10551-012-1574-1

De Graaf, F. J., & Stoelhorst, J. W. (2013). The role of governance in corporate social responsibility: Lessons from Dutch finance. *Business and Society, 52*(2), 282–317.

Deakin, S., & Hobbs, R. (2007). False dawn for CSR? Shifts in regulatory policy and the response of the corporate and financial sectors in Britain. *Corporate Governance: An International Review, 15*(1), 68–76.

Edmans, A. (2011). Does the stock market fully value intangibles? Employee satisfaction and equity prices. *Journal of Financial Economics, 101*(3), 621–640.

Faleye, O., & Trahan, E. A. (2011). Labor-friendly corporate practices: Is what is good for employees good for shareholders? *Journal of Business Ethics, 101*(1), 1–27.

Girerd-Potin, I., Jimenez-Garces, S., & Louve, P. (2014). Which dimensions of social responsibility concern financial investors? *Journal of Business Ethics, 121*(4), 559–576.

Gompers, P. A., & Metrick, A. (2001). Institutional investors and equity prices. *The Quarterly Journal of Economics, 116*(1), 229–259.

Goss, A., & Roberts, G. S. (2011). The impact of corporate social responsibility on the cost of bank loan. *Journal of Banking and Finance, 35*(7), 1794–1810.

Graves, S. B., & Waddock, S. A. (1994). Institutional ownership and corporate social performance. *Academy of Management Journal, 37*(4), 1034–1046.

Ghoul, S. E., Guedhami, O., Kwok, C. Y., & Mishra, D. R. (2011). Does corporate social responsibility affect the cost of capital? *Journal of Banking and Finance, 35*(9), 2388–2406.

Harrison, J. S., & Freeman, R. E. (1999). Stakeholders, social responsibility, and performance: Empirical evidence and theoretical perspectives. *Academy of Management Journal, 42*(5), 479–485.

Johnson, R. A., & Greening, D. W. (1999). The effect of corporate governance and institutional ownership types on corporate social performance. *Academy of Management Journal, 42*(5), 564–576.

Kang, J. K., & Stulz, R. M. (1997). Why is there a home bias? An analysis of foreign portfolio equity ownership in Japan. *Journal of Financial Economics, 46*(1), 3–28.

Kho, B. C., Stulz, R. M., & Warnock, F. E. (2006). *Financial globalization, governance, and the evolution of the home bias*. NBER Working Paper Series 12389.

Leuz, C., Lins, K. V., & Warnock, F. E. (2009). Do foreigners invest in poorly governed firms? *The Review of Financial Studies, 23*(3), 3245–3285.

Miyajima, H., & Nitta, K. (2011). *Diversification of shareholding structure and the results: Dissolution and revive of cross-shareholdings and role of foreign investors*. RIETI Discussion Paper Series 11-J-011 (in Japanese).

Neubaum, D. O., & Zahra, S. A. (2006). Institutional ownership and corporate social performance: The moderating effects of investment horizon, activism, and coordination. *Journal of Management, 32*(1), 108–131.

Petersen, M. A. (2009). Estimating standard errors in finance panel data sets: Comparing approaches. *Review of Financial Studies, 22*(1), 435–480.

Scholtens, B., & Zhou, Y. (2008). Stakeholder relations and financial performance. *Sustainable Development, 16*(3), 213–232.

Stewart, F. and Yermo, J. (2010). Options to improve the governance and investment of Japan's government pension investment fund. *OECD Working Papers on Finance, Insurance and Private Pensions.*

Surroca, J. J., Tribo, A., & Waddock, S. (2010). Corporate responsibility and financial performance: The role of intangible resources. *Strategic Management Journal, 31*(5), 463–490.

Suto, M., & Toshino, M. (2005). Behavioral biases of Japanese institutional investors: Fund management and corporate governance. *Corporate Governance: An International Review, 13*(4), 466–477.

Turban, D. B., & Greening, D. W. (1997). Corporate social performance and organizational attractiveness to prospective employers. *Academy of Management Journal, 40*(3), 658–672.

Impacting Factors of Triple Performance of Farmer's Professional Cooperatives in China: A Case Study of Jiangsu Province

Hualiang Lu, Dongqin Wang, and Mengya Wang

1 Introduction

Cooperatives first originated from Europe. In China, the farmers' professional cooperatives began to sprout in 1980s, and had entered a rapid development stage in the twenty-first century. For the study of cooperatives, foreign scholars wanted to know which one has more advantages for farmers or investor owned farmers (IOF). Eventually they agreed that cooperative can better safeguard the interests of farmers. Of course, comparing with the IOF, the cooperative organization has many defects. First, since principal-agent relationship exists in cooperatives, the governance cost is very high and if managers can't do a good job, the members of cooperatives could cause lazy behavior easily. Second, as an organization of collective ownership, cooperatives have low efficiency in resource allocation. Finally, the control costs of cooperatives will also increase with the enlargement of scale. So the size of cooperatives is not able to get optimization either (Porter and scully 1987).

In recent years, the research on the role of farmers' professional cooperatives is mainly concentrated on the influence of farmers' professional cooperatives to farmers' welfare. The more consistent view of scholars is that the establishment of farmers' professional cooperatives can improve farmers' income. Studies found that farmers who joined cooperatives could gain higher transaction price by selling products through the professional market than through traditional market, so the economic performance of farmers could be improved (Costales 2002). At the same time, cooperatives can improving operation performance by guarantee farmers to get better production materials and services (Maharjan and Fradejas 2006; Toro and Hansson 2004). Chinese study on the role of farmers' professional cooperatives

H. Lu (✉) • D. Wang • M. Wang
Sino-German Research Institute for Sustainable Development, Nanjing University of Finance and Economics, Nanjing, China
e-mail: hualianglu@nufe.edu.cn; 823793265@qq.com; 512740825@qq.com

H. Lu et al. (eds.), *Building New Bridges Between Business and Society*, CSR, Sustainability, Ethics & Governance, DOI 10.1007/978-3-319-63561-3_12

mainly focused on the influence of cooperatives on ensuring quality and safety of agricultural products and promoting farmers' income. Many Chinese scholars make a theoretical analysis of farmers' professional cooperatives from the perspective of farmers' income. They concluded that the development of farmers' professional cooperative can improve the degree of farmers' organization and agricultural commercialization, and will foster industrialization of agriculture and promote farmers' income (Chen 2005; Wang 2001; Kong et al. 2005; Du 2005; Cai 2011).

For cooperative governance, Shao and Xu (2008) analyze it in two aspects. One is in the established organization environment and the ownership structure, members of cooperative could choose who have control rights, especially the control of contingency. The other one is how members control the owner of control right. As an economic organization with multiple characteristics, cooperative governance structure could be explained by principle-agent theory and transaction cost theory.

From the perspective of principle-agent theory, cooperative is a collection of agents who provide the resource to economic production activities and then get different kinds of economic interests. Zhang (1995) argues that on the premise of clearing the principal and agent, reasonable incentive system can achieve the purpose of controlling agent. Ma (2006) proposed that cooperative has dual principle-agent relationships. One is the traditional principle-agent relationship between all members and operators. The other one is between the small-medium-sized members and the core members. And the second one is the main aspect of contradiction. Therefore we should consider the principle-agent problem from the aspects of cooperative internal supervision, organizational incentive and so on. And we must take more consideration on multiple principle-agent relationships.

Scholars in 1980s used transaction cost theory to explain the development of cooperatives. Levay (1983) argued that cooperatives could provide a path to enter the market, keep the stability of the market in a long term, and gain positive benefit by achieving economies of scale. Cooperatives could also reduce production cost and transaction cost, and increase membership income. Demsetz argues that comparing with other organizations, costs of cooperatives are much higher. One possible reason is that cooperatives must pay higher supervision cost in order to reduce opportunism behavior. So clear property rights is the key to solve the sustainable development of cooperatives.

Although performance evaluation system for enterprises are well developed, such as the balanced scorecard, economic value added system, performance triangular prism system. But the performance evaluation for cooperatives is still lack of effective evaluation system. The "Triple Bottom Line" theory was introduced in 1998 by John Elkington, which refers that we need to consider the balance development of economic, environment and social welfare when the enterprises pursuit of their own development. Hendrikse and Veerman (2001) pointed out that farmers' professional cooperatives have for cooperatives double identities of member's community and enterprise. Therefore the economic and social functions of the cooperatives should be equally considered. Gertler (2001) argued that the economic, social and ecological performance must be fully considered for cooperatives.

This paper taking farmers' professional cooperatives in Jiangsu province as an example, develops a comprehensive analytical research framework to study the key factors affecting the performance of farmers' professional cooperatives.

2 Theoretical Background and Research Model

Based on principal-agent theory, transaction cost theory and triple performance theory, this paper mainly studies the influence of internal and external environment on cooperative governance mechanism and on the economic, social, environmental performance of cooperatives in China.

2.1 The Influence of External Environment on Governance Structure and Cooperative Performance

Farmers' professional cooperatives are developed under the supervision and regulation of government. The state shall encourage and support the development of cooperatives. Cooperatives are regarded as the no-profit economic organization which will serve the farmers. So cooperatives in the process of governance have got a lot of influence of external environment. The supervision and management of government will affect the arrangement of cooperative governance structure. And this will further affect cooperatives' performance. The reason that causes the unreasonable cooperative internal governance structure and irregular management system is the tight external operation environment and multiple management of government. All of these will affect the healthy operation of farmers' professional cooperatives. Institutional environmental factor, the government support and standardize management are the key factors which will affect farmers' professional cooperative performance (Huang et al. 2002).

From the previous research, we can find that the operation of farmers' professional cooperative is influenced by a variety of external factors, such as the construction of legal system, government support, technical talented person, members quality, industrialization management and the level of organizational management (Chang et al. 2009). Meanwhile, Zhao and Jiang (2009), Fulton and Sanderson (2002) pointed out that some external factors will affect cooperative performance, such as the perfection degree of cooperative laws, support degree for cooperative, and the development of cooperatives industrialization. At the same time, external environment largely influences willingness of participation of farmers' professional cooperative. And then the development of local industry and the level of economic development will also influence performance of the farmers' professional cooperatives. Huang et al. (2002) also stressed that cooperative's own financing ability and financing environment could influence the development of cooperatives. Feinerman and Falkovitz (1991) pointed out that only cooperative timely pay attention to the change of external economic environment, can realize their sustainable management.

2.2 The Influence of Internal Environment on Governance Structure and Cooperative Performance

Heterogeneity of members in farmer' professional cooperatives refers to a kind of feature differentiation between members which is different from the features of members in traditional cooperatives. Its basic performance is the difference of pursuing interests between members. So the heterogeneity of members will have a significant impact on the cooperative governance structure (Shao and Xu 2008). Due to the existence of membership heterogeneity in cooperatives, the democratic management of one person one vote must be applied. And it should be ensure the relative homogeneity of the council and the board of supervisors, and realize effective balances of the two boards. Many research results have shown that heterogeneity of cooperative members has a significant impact on the development of farmers' professional cooperatives (Huang et al. 2002, 2007, 2008; Zhang 2008). Farmers' professional cooperatives operating different kinds of products have different income, surplus, and operation management (Huang and Shao 2010). So policy support for farmers' professional cooperatives and performance evaluation must be considered based on the differentiation of product features. We regard the cooperative members' heterogeneity and cooperative with different kinds of product features as the internal environment of cooperative. As a result, the cooperative's internal environment will affect the cooperative governance structure. Chang et al. (2009) suggested that the key internal factors of farmers' professional cooperatives are: organization member quality, industry characteristics and industrialization. Zhang et al. (2010) showed that the age, level of education of the main managers in cooperatives and organization scale have significant impact on performance.

2.3 The Influence of Governance Structure on Cooperative Performance

Previous studies have shown that cooperative governance structure is the core factor which affect cooperative performance (Viltaliano 1993; Cook 1995; Nilsson 2001). Viltaliano (1993) argued that the way carried out in accordance with the volume return surplus, and governance structure of one-people-one-vote would be helpful to the development of cooperatives. Because of the difference of the responsibility that members should take, there are core members and ordinary members in the farmers' professional cooperatives. Core members often refer to the sponsors, leaders and large stakeholders. And the ownership concentration, personal quality, academic level and social network of core members will have a significant impact on the development of cooperative performance (Huang et al. 2002).

Huang (2007) argued that the cooperative performance was influenced by cooperative governance structure in a large extent. Using date from the survey of 168 farmers' professional cooperatives in Tai Zhou, Wen Zhou and Han Dan, the results showed that the performance closely related to the governance mechanism. Because of the existence of double agency problem in farmers' professional cooperatives and the high transaction cost, the cooperative governance structure, especially under the condition of present heterogeneity of cooperative members, will have a significant impact on cooperative performance.

2.4 The Influence of Production Operations to Cooperative Performance

The farmers' professional cooperatives in China came out for coping with the defects of decentralized management model and the changeable market economy. Therefore the organizational service function is an important indicator to measure the cooperative performance. Previous studies showed that farmers' professional cooperative in China played many functions in the process of production operation. Huang et al. (2010) divided the productive service of farmers' professional cooperative into four types, including providing information technology or service, agricultural material, agricultural products sales and finance credit services. Here, we focus on three aspects of the production operation of farmers' professional cooperative: unity function of cooperatives, production flexibility of cooperatives and cooperative's services.

On the basis of previous studies, we try to analyze the influence of international environment, external environment on governance structure, and production operation on cooperative performance (Fig. 1).

3 Measurement

3.1 Sample Selection and Data Collection

The samples of this research are farmers' professional cooperatives within the scope of Jiangsu Province. And interviewees are the heads of farmers' professional cooperative who are very familiar with the cooperative production and operation (including the director, supervisor, boss, etc.). The sample selection is based on the

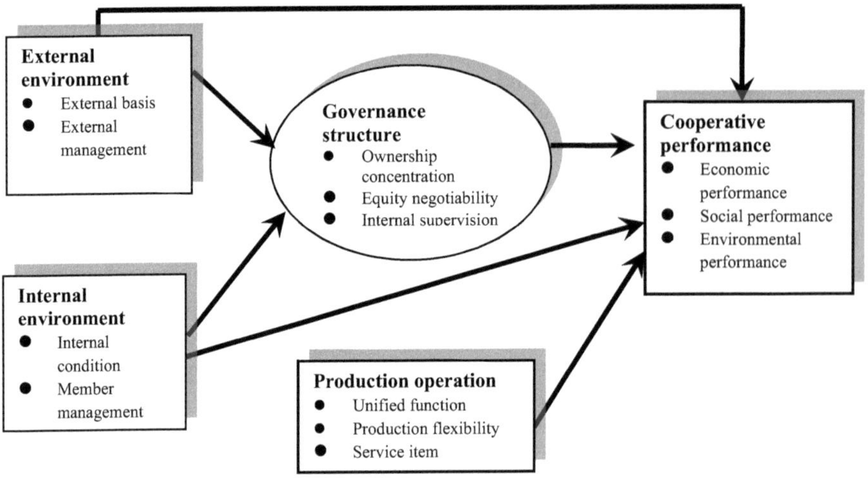

Fig. 1 Theoretical framework models. Source: Huang et al. (2002), Lin and Huang (2007) and Huang and Xu (2008)

average area yield of grain in every county in the last 3 years, and then we randomly select sample counties. First we determine the requirement of extracting 18 counties, and then make sure the first sample by a random number. The other samples are extracted according to certain interval (five counties). Finally we extract 18 counties which are distributed in the 11 cities except Wuxi and Changzhou.

We have one-on-one interviews with the heads of farmers' professional cooperatives. Through this survey, we get 112 questionnaires. Because of the missing data in three cooperatives, and we eliminate these three questionnaires. And there are 34 cooperatives that don't exist stock. Therefore in this paper, there are 75 valid questionnaires, accounts for 67.0% of total questionnaires.

3.2 *Methods*

This questionnaire is mainly composed of six parts. The first part is descriptive questions about the basic conditions of farmers' professional cooperatives, and also includes some messages of director and sponsor. The second part is about the internal governance structure of farmers' professional cooperatives. The third, fourth and fifth parts are the cooperative internal and external environment, production operation situation, the contribution to the economic, society and environment, internal governance mechanism and supervision respectively. We adopt the form of Likert 5 scales to measure these three parts. And 1 = completely disagree, 2 = not agree, 3 = neutral, 4 = agree, 5 = completely agree. Respondents (usually the head of farmers' professional cooperative) can score these questions according to the actual situation of the farmers' professional cooperatives.

According to review of above theory, we will use external basis and external supervision to measure external environment. The external basis of cooperative is to measure whether it is benefit to the development of cooperative. This variable is measured by four items, including financing situation, industrial development, the cooperation relationship with leading enterprises and the development situation of production base. External supervision can be measured in two items. The first one is the government has strict supervision and regulation to cooperative. Second one is the current policy of government promotes the development of cooperatives.

Internal environment of cooperatives can be classified into internal conditions and member management. Internal conditions can be measured by four items, while member management is measured by two items, including "members have strong restriction to quit the cooperatives" and "members have strong restriction to join cooperatives".

The governance structure of farmers' professional cooperatives is measure by three variables: ownership concentration, equity negotiability and internal supervision. Ownership concentration is measured by four items. Equity negotiability is measured by two items, which are "shares can be transferred free" and "shares can be transferred to the members of non-cooperatives". The internal supervision is measured by four items. These four items reflects the internal supervision mechanism of farmers' professional cooperatives.

The unity of production operation is measured by three variables: unity function, production flexibility and service content. Unity function and production flexibility are measured respectively by two items, while service content is measured by four items.

Cooperative performance is measured by economic, social and environmental performance.

Economic performance is mainly measured by financial indicators, social performance is mainly measured by non-financial indicators, and environmental performance is measured by some indicators related to the environment. Specifically, the measurement of economic performance mainly adopts cooperative production scale and its value-added (such as the total number of members, cooperative fixed assets, cooperative annual income) and management benefit (such as cooperative surplus, etc.). And the measurement of social performance is mainly to measure the contribution that cooperative can contribute to the rural development (such as the number of local farmers and service providing for members). For the evaluation of environmental performance, we can mainly consider the cognitive situation of pollution-free agricultural products, green agricultural products or organic agricultural products, and the use of environmental protection pesticides and fertilizers, and the recycle of waste etc.

4 Data Analysis and Results

4.1 Reliability and Validity Analysis

The first step of data analysis is to test the reliability and validity. First, we use reliability analysis to test the reliability, authenticity, consistency and stability of the research data. In this paper, we mainly use Cronbach's Alpha, average variance extracted and composite reliability to test the reliability. These measurement values are acquired by SPSS 17.0 and PLS 2.0 analysis software. Through the reliability test, we can find that the Cronbach's Alpha of all variables are above 0.6. The composite reliability of all variables is above 0.8, AVE values are bigger than 0.5. These show that the data can be further analyzed and discussed to ensure the accuracy of analysis results.

In this study, we use factor analysis to analyze the structure validity, and use the KMO and Bartlett sphericity test to determine whether the data in the questionnaire is suitable for factor analysis.

It shows that the KMO value of all the variables is bigger than 0.5, and the P value of Bartlett sphericity test is less than 0.001. And the results show that all these variables are suitable for factor analysis.

4.2 Correlation Analysis

Table 1 is a Pearson correlation coefficient matrix of all factors. As we can see from Table 1, farmers' professional cooperatives are closely related to the internal and external environment, the production operation and governance structure etc. It also lays a foundation for our further study.

4.3 Structural Equation Modelling Analysis

We use Partial Least Squares to give an empirical analysis on the sample data. In the structural equation model, we not only study the impact of internal and external environment of cooperative on the governance structure, which will have an indirect effects on triple performance, but also consider the direct influence of internal and external environment of cooperatives on triple performance.

(1) The internal and external environment of cooperatives will impact on governance structure.

Figure 2 shows that the external basis of cooperative has a significant impact on internal supervision and equity negotiability. That is to say, under the condition of well financing situation, well local industry development and good relationship between cooperatives and leading enterprises, the frequency of cooperative

Table 1 Correlation matrix of variables

	1	2	3	4	5	6	7	8	9	10	11	12	13
1. External basis	1												
2. External management	**0.268****	1											
3. Internal condition	**0.287****	0.132	1										
4. Member management	0.069	**0.212****	**0.214****	1									
5. Service items	**0.529****	**0.292****	**0.246**	0.069	1								
6. Production flexibility	0.142	−0.049	**0.281****	0.102	0.126	1							
7. Unity function	0.057	0.042	0.172	**−0.23****	0.145	**0.359****	1						
8. Equity negotiability	−0.154	−0.147	−0.091	−0.057	−0.065	−0.119	0.131	1					
9. Ownership concentration	0.038	**−0.193****	0.058	−0.072	0.077	0.111	0.145	0.078	1				
10. Internal supervision	0.095	**0.321****	**0.439****	**0.320****	**0.292****	0.071	0.073	**−0.186****	0.058	1			
11. Environmental performance	0.111	0.022	0.079	**−0.236****	**0.246****	0.157	**0.327****	0.081	0.174	−0.104	1		
12. Social performance	**0.485****	**0.366****	**0.549****	0.141	**0.502****	**0.431****	0.093	**−0.261****	−0.079	**0.329**	**0.238**	1	
13. Economic performance	**0.467****	**0.401****	**0.342****	0.154	**0.328****	−0.069	0.006	−0.164	−0.118	**0.282**	−0.057	**0.436**	1

Note: ** refers to the correlation coefficient is significant at 5%.

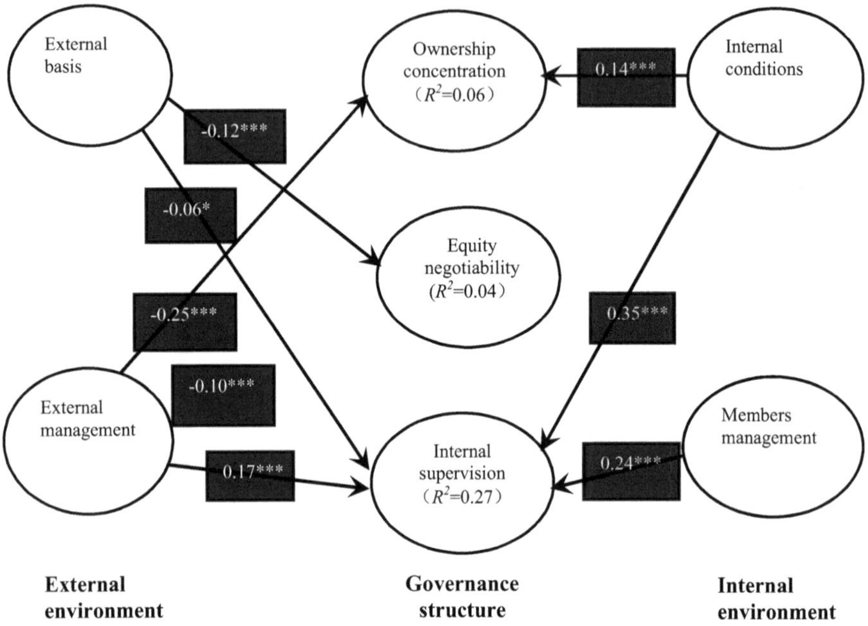

Fig. 2 Influence of internal and external environment on governance structure

congress convened and the board of supervisors will decrease, also the time of annual financial disclosure and the number of members in the board of supervisors will decrease, and the shares of members at the same time will tend to be free transferred. External management has significant negative impact on ownership concentration. It shows that the more supervision and regulation of agricultural administrative department to cooperative, the better development of current policy of government to cooperative, and the less ownership concentration cooperative has. Internal conditions and members management of cooperatives have significant positive impact on internal supervision. It shows that if cooperatives have good offices and service facilities cooperatives that hold members meeting and conduct a comprehensive record will strengthen internal supervision.

(2) Governance structure and production operation will impact on triple performance.

From Fig. 3, we can see that equity negotiability only has significant negative influence on social performance. Maybe because if equity can be free transferred, it will be not benefit for the management of members, then it can't guarantee the increase income of farmers who join the cooperative. The situation that a few people have equity is not conductive to improve economic performance. And this conclusion is opposite to the results of Wu and Xu (2013). They argue that the high ownership concentration of the core members will have a positive impact on the development and profitability of cooperative. Maybe because of they regard the cooperative performance as a comprehensive performance. While we analyze the performance from three aspects: economic, social and environmental performance.

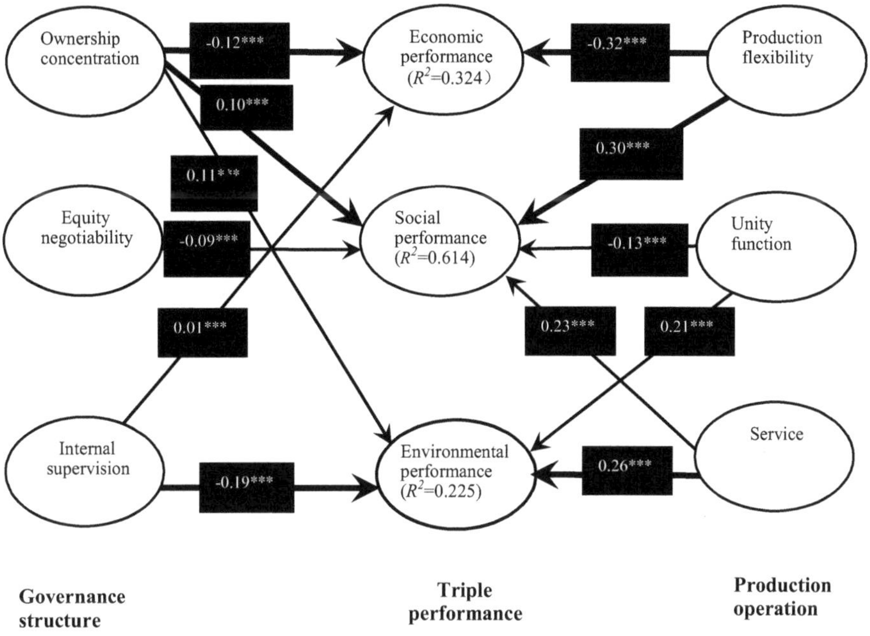

Fig. 3 Influence of governance structure and production operation on triple performance

The influence of internal supervision on social performance is not very obvious. This result is similar to the result of Wu and Xu (2013). They point out that the influence of internal supervision on the overall performance only accounts for 5.41%, which is not particularly significant.

In the process of production operation, the unity function and service items of cooperatives don't have significant impact on economic performance. But their influence on social performance is significant. Production flexibility has negative impact on economic performance. And this shows that changing size arbitrarily will change the stability of the cooperatives' operating income in a great degree. Thus production flexibility presents the negative effect to economic performance.

(3) Total effect analysis

From Table 2, we can see that most of the paths are significant. But the external basis has no significant effect on environmental performance and ownership concentration. External management and equity negotiability also have no obvious significant effect on environmental performance. Internal conditions and member management have no significant effect on equity negotiability. Member management, service items, equity negotiability and unity functions also have no significant effect on economic performance. At the same time, internal supervision has no significant effect on social performance.

From Table 2 we can see the influence of each variable on triple performance. The external basis and external management have significant effect on economic and social performance at the level of 1%. This is consistent with the results of Feinerman and Falkovitz (1991) and Egerstorm (2004). But the external environment

Table 2 Total effects

	Estimated value of PLS	Mean	Standard deviation	T-value	Sig. level
External basis → internal supervision	−0.063	−0.068	0.040	1.649	*
External basis → environmental performance	−0.036	−0.030	0.051	1.074	
External basis → social performance	0.204	0.173	0.032	5.363	***
External basis → economic performance	0.351	0.357	0.043	8.429	***
External basis → equity negotiability	−0.121	−0.117	0.052	2.294	**
External basis → ownership concentration	0.071	0.071	0.052	1.325	
External management → internal supervision	0.169	0.172	0.045	3.773	***
External management → environmental performance	−0.014	−0.016	0.036	1.357	
External management → social performance	0.192	0.165	0.033	3.571	***
External management → economic performance	0.162	0.199	0.038	4.044	***
External management → equity negotiability	−0.103	−0.108	0.048	2.150	**
External management → ownership concentration	−0.250	−0.248	0.037	6.653	***
Internal conditions → equity negotiability	−0.016	−0.017	0.044	0.346	
Internal conditions → ownership concentration	0.138	0.136	0.055	2.485	**
Internal conditions → internal supervision	0.352	0.354	0.036	9.774	***
Internal conditions → environmental performance	0.023	0.023	0.044	1.664	*
Internal conditions → social performance	0.315	0.313	0.044	7.027	***
Internal conditions → economic performance	0.091	0.190	0.051	3.298	***
Members management → internal supervision	0.214	0.211	0.045	4.718	***
Members management → equity negotiability	−0.212	−0.023	0.049	0.479	
Members management → ownership concentration	−0.054	−0.052	0.041	1.296	
Members management → environmental performance	0.117	−0.211	0.039	4.158	***
Members management → social performance	−0.023	−0.054	0.029	2.651	***
Members management → economic performance	−0.053	0.073	0.038	1.163	

(continued)

Table 2 (continued)

	Estimated value of PLS	Mean	Standard deviation	T-value	Sig. level
Service items → environmental performance	0.263	0.257	0.068	3.868	***
Service items → social performance	0.229	0.272	0.040	6.793	***
Service items → economic performance	0.039	0.040	0.049	0.801	
Production flexibility → environmental performance	0.060	0.052	0.045	1.280	
Production flexibility → social performance	0.298	0.339	0.036	9.432	***
Production flexibility → economic performance	−0.316	−0.187	0.055	3.404	***
Unity function → environmental performance	0.210	0.215	0.043	4.877	***
Unity function → social performance	−0.125	0.125	0.028	4.491	***
Unity function → economic performance	0.099	0.037	0.040	0.917	
Equity negotiability → environmental performance	0.031	0.034	0.055	0.538	
Equity negotiability → social performance	−0.092	−0.099	0.030	3.270	***
Equity negotiability → economic performance	−0.070	−0.068	0.047	1.429	
Ownership concentration → environmental performance	0.114	0.109	0.040	2.793	***
Ownership concentration → social performance	−0.103	−0.119	0.022	5.563	***
Ownership concentration → economic performance	−0.122	−0.092	0.035	2.684	***
Internal supervision → environmental performance	−0.187	−0.189	0.035	2.684	***
Internal supervision → social performance	0.099	0.053	0.033	1.620	
Internal supervision → economic performance	0.013	0.100	0.044	2.264	**

Note: ***means $p < 0.01$; **means $p < 0.05$; *means $p < 0.10$

has no significant effect on environmental performance. One possible reason is that environmental performance may have more effect on internal environment of cooperatives. Decreasing the use of chemical fertilizers, recycling the waster as well as ensuring the safety of agricultural product have closely relevant to members management. The participating and exiting of corresponding members did not affect the cooperative economic performance.

Through Table 2, we can see that internal conditions have greatest effect on social performance. This is because the internal environment has a great effect on internal supervision, and then it has a certain influence on social performance through internal supervision. At the same time, it also has a direct effect on social performance (Guo et al. 2009).

5 Conclusions and Implications

This paper uses structural equation modeling to explorer the key factors which influence the triple performance of cooperatives in China. We had field interview and collected questionnaire survey for 75 sample cooperatives in 11 areas of Jiangsu Province.

The results shows that ownership concentration, equity negotiability and production flexibility have significant negative impact on economic performance; Membership management, equity negotiability and ownership concentration have significant negative impact on social performance; Member management and internal supervision have significant negative effect on environmental performance. This shows that only ease restrictions on farmers participating and exiting can contribute to the environmental performance.

From the critical paths affecting triple performance of agricultural cooperatives in China in Fig. 4, we can see that if we want to improve the economic performance of cooperatives, we must pay much attention on the influence of

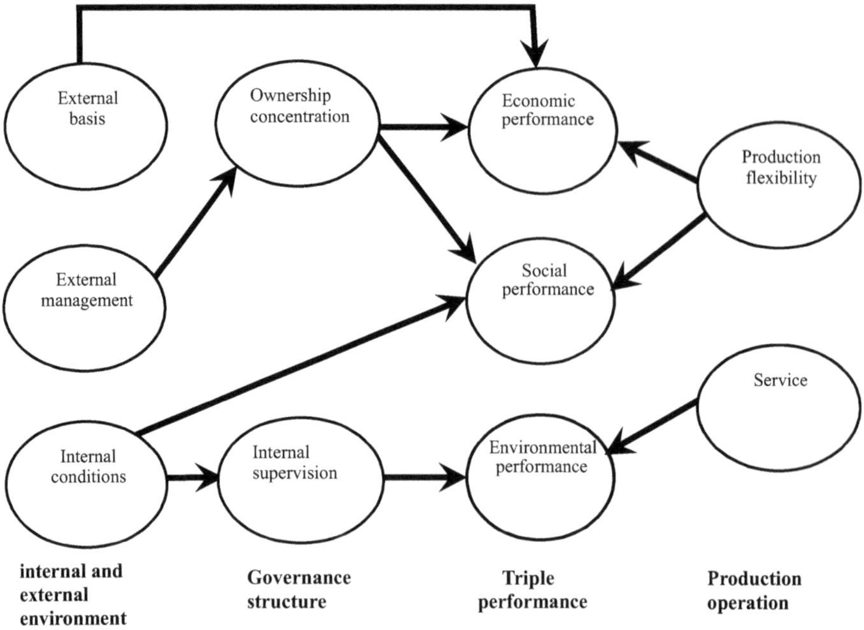

Fig. 4 Critical paths affecting performance

external management on ownership concentration of governance structure, and take a direct consideration of the local industry development and relationship with leading enterprises. Focusing on internal conditions, ownership concentration and production flexibility can improve social performance. And also pay special attention on improving internal conditions, strengthen internal supervision and service items provided by cooperative members in order to improve environmental performance of cooperatives in Chinese context.

According to the above analysis of the development of farmers' professional cooperatives in Jiangsu Province, we put forward following management suggestions: (1) improve external supervision and management mechanism for leading farmers' professional cooperatives; (2) pay attention to combining with the local industry development, and strengthen the contact with leading enterprise; (3) improve the internal environment of cooperatives; (4) reduce ownership concentration, and improve internal supervision; (5) improve service function of farmers' professional cooperatives, and expand their production capacity.

Acknowledgements We gratefully acknowledge the financial support from WOTRO sponsored project "Cooperatives and Chains", the National Natural Science Foundation of China (71173097, 711773046), "Youth Talent Project" and "Six Talent Peak Project" of Jiangsu Province.

References

Cai, R. (2011). The model of "cooperative+farmer": Transaction cost saving and farmers' income effect—The empirical analysis of questionnaire based on the survey of apple planting farmers in Shan Dong province. *Chinese Rural Economy*, (1), 58–65.

Chang, W., Hu, H., & Yang, S. (2009). Relational structure analysis: The study of influence factors of economic organization operation of farmers' professional cooperatives in China. *Journal of Shanghai Economic Studies, 1*, 23–29.

Chen, S. (2005). How to enliven the circulation of agricultural products, promote the farmers' income. *Journal of Cooperative Economy in China, 6*, 7–11.

Cook, M. L. (1995). The future of U.S. agricultural cooperatives: A neo-institutional approach. *American Journal of Agricultural Economics, 77*, 1153–1159.

Costales, A. C. (2002). *Market and policy changes impacting on smallholder hog producer's participation in the growing Philippine market*. Professional Chair Lecture, College of Economics and Management, University of the Philippines Los Banos (Uplb), Laguna, Philippines.

Du, Y. (2005). The impact of industrialized operation of agriculture and innovation of farmers' organization on farmers' income. *Chinese Rural Observation*, (3), 9–17.

Egerstorm, L. (2004). Obstacles to cooperation. In C.D. Merrett & N. Walzer (Eds.), *Cooperatives and local development: Theory and applications for the 21st century* (pp. 70–91). Routledge.

Feinerman, E., & Falkovitz, M. S. (1991). An agricultural multipurpose service cooperative: Pareto optimality, price-tax solution, and stability. *Journal of Comparative Economics, 15*, 95–114.

Fulton, M., & Sanderson, K. (2002, March). *Co-operatives and farmers in the new agriculture*. Report prepared for the co-operatives secretariat agriculture and Agri-food Canada. Saskatoon, Canada: University of Saskatchewan, Center for the Study of Co-operatives.

Gertler, M. (2001). *Rural cooperatives and sustainable development*. In center for the study of cooperative University of Saskatchewan, C. c. i. P. d., printed in Canada, http://coop-studies.usask.ca

Guo, H., Lou, D., Hu, Z., et al. (2009). The factors analysis that influence the growth of farmers' professional cooperative—Based on the survey of part cooperatives in Zhejiang province. *Chinese Rural Economy*, (8), 24–31.

Hendrikse, G. W. J., & Veerman, C. P. (2001). Marketing cooperatives: An incomplete contracting perspective. *Journal of Agricultural Economics, 52*(1), 53–64.

Huang, S. (2007). *The organization behavior study of farmers' professional cooperatives in the transformation time: Based on the perspective of member heterogeneity*. Hangzhou: Zhejiang University.

Huang, J., Deng, H., & Xu, Z. (2010). The study of service function of Chinese farmers' specialized economic cooperative organization and its influencing factors. *Management World*, (5), 75–81.

Huang, S., & Xu, X. (2008). Study on cooperatives' membership hetrogeneity and organizational structure. *Journal of Nanjing Agricultural University (Social Sciense Edition), 8*(3), 1–7.

Huang, Z., Xu, X., & Feng, G. (2002). Study on the impacting factors of the development of farmers'professional cooperatives. *China Rural Economics, 3*, 13–21.

Huang, S., Lin, J., & Xu, X. (2008). The empirical analysis of the governance mechanism and performance of farmers' professional cooperatives. *Chinese Rural Economy, 279*(3), 65–73.

Huang, Z., & Shao, K. (2010). The analysis of organizational structure of farmers' professional cooperatives and operating performance based on the features of product. *Journal of Academic Exchange, 196*(7), 91–96.

Huang, Z., Xu, X., & Feng, G. (2002). The influence factors analysis of farmers' professional cooperative—Discussed the present development of farmers' professional cooperative organization in Zhejiang province. *Chinese Rural Economy*, (3), 13–21.

Kong, X., Zhang, X., & Pang, X. (2005). The survey of the role and restricting factors of farmers' cooperative organization in Shan, Ning, and Chuan area. *Journal of Economic Theory and Economic Management, 6*, 52–57.

Levay, C. (1983). Agricultural cooperative theory: A review. *Journal of Agricultural Economics, 34*(1), 1–44.

Lin, J., & Huang, S. (2007). Study on cooperatives' membership hetrogeneity and ownership. *Agricultural Ecnonomic Issues, 10*, 12–17.

Ma, Y. (2006). *The system resolution of farmers' professional cooperatives in China—Based on Zhejiang Province*. PhD thesis of Zhejiang University.

Maharjan, K. L., & Fradejas, C. C. (2006). *Role of cooperative in proving access to production resources and household economy of Baekyard Pig Raisers in Batangas, Philippines* (Vol. 8, pp. 1–16). Poster Paper Prepared for Presentation at the International Association of Agricultural Economists Conference. Gold Coast, Queensland, Australia.

Nilsson, J. (2001). Organizational principles for cooperative firms. *Scandinavian Journal of Management, 17*, 329–356.

Porter, P. K., & Scully, G. W. (1987). Economic efficiency in cooperatives. *Journal of Law and Economics, 30*, 489–512.

Shao, K., & Xu, X. (2008). Member heterogeneity of farmers' professional cooperative has impact on cooperative governance structure—Based on the analysis of 88 cooperatives in Zhejiang Province. *Journal of Northwest Agriculture and Forestry University of Science and Technology (Social Science Edition), 8*(2), 5–9.

Toro, A. D., & Hansson, P. A. (2004). Machinery cooperatives: A case study in Sweden. *Biosystems Engineering, 87*(1), 13–25.

Viltaliano, P. (1993). Cooperative enterprises: Alternative conceptual basis of analysis a complex institution. *American Journal of Agricultural Economics, 65*, 1078–1083.

Wang, W. (2001). The farmers' cooperative economy and increasing farmers' income. *Journal of Agricultural Economy*, (9), 23–24.

Wu, B., & Xu, X. (2013). The study of cooperative state characteristics which impact on the governance structure—Based on the survey of 266 farmers' professional cooperatives in 80 counties of 3 Provinces in China. *Agricultural Technology Economy, 1*, 107–119.

Zhang, J. (2008). The analysis of heterogeneity of the internal members in farmers' professional cooperatives. *Journal of Anhui Agricultural Science, 4*(29), 12919–12920.

Zhang, B., Ou, G., Yao, M., et al. (2010). *The study of economic organizational development of farmers' professional cooperatives—In the case of Jiangsu province*. Beijing: China Agriculture Press.

Zhang, W. (1995). *Enterprise enterpreneurs—A contract theory* (pp. 1–39, 5–216). Shanghai: The joint publishing company LTD of Shanghai.

Zhao, J., & Jiang, T. (2009). A triple performance evaluation model of farmers' professional cooperatives. *Journal of Hunan Agricultural University (Social Science Edition), 10*(4), 1–7, 26.

Integrative Model in Mitigating the Impact of International Labor Migration on Family Left Behind: Case Study in Indramayu District, Indonesia

Lina Widyastuti

1 Introduction

Over recent decades, the number of international labour migrants is rapidly increasing throughout the world. This phenomenon has been high on the global agenda with increasing emphasis on migration and development linkages. The UN High Level Dialogue on International Migration and Development (New York, 14–-15 September 2006) was a milestone reflecting this trend. The major portion of these migrants belong to developing countries including Indonesia. However, it is unfortunately not possible to provide a totally accurate picture of international labor migration movement from Indonesia because the statistics available exclude undocumented workers. For legal labour migration, The National Agency for the Placement and Protection of Indonesian Migrant Workers (BNP2TKI) mentioned that there were 152,022 Indonesian workers stationed abroad in the period of 2006–2013, of which 58% are women. The principal destination countries include Middle East, Singapore, Malaysia and Hong Kong.

The recent discourse on international labour migration highlighted the role of remittances, return migration and circulation, and transitional communities as major factors in promoting development in countries of origin. Migration has always been positively viewed in terms of the visible monetary gains generated for the original countries of the labours. The massive amounts of remittances flowing into national coffers is the main reason these countries are institutionalizing market-and demand-driven migration policies to further encourage the labour migration of hundreds of thousands of women and men to different countries (UN Women 2013). Furthermore, there is the possibility that remittances sent

L. Widyastuti (✉)
Directorate of Population Impact Analysis, The Indonesia National Population and Family Planning Board (BKKBN), Jakarta, Indonesia
e-mail: lina_whe@yahoo.com

H. Lu et al. (eds.), *Building New Bridges Between Business and Society*, CSR, Sustainability, Ethics & Governance, DOI 10.1007/978-3-319-63561-3_13

from abroad will relax the household budget constraint. Several studies found evidences supporting this hypothesis (Alcaraz et al. 2012; Yang 2008). Nevertheless, the issue of social costs of migration especially its impact on families left behind remains largely unrecognized, understudied and politically less favorable. Migration inevitably not only affects the individual migrants, but also families and communities in many different ways (Antman 2012). For instance, parental migration inherently involves parental absence from the house which may have negative impacts over their children that can outweigh the positive effect of remittances.

The family is the first and foremost institution in the task of meeting the needs of growth and development of children. In the family in which one member of his family became migrant workers overseas, the possibility of family functions can be disrupted. The direction and magnitude of these effects, however, are yet fully understood. Indramayu is one of the largest sending migrants district in Indonesia. However, studies on the socio-cultural and economic impacts of overseas migration on emigrants families left behind in the rural areas of Indramayu are scarce. This study aimed to evaluate the impact of international labour migration on family structures, the condition of their left-behind children, saving and investment patterns of the migrants' families left behind. The study also developed a strategic solution model based on its findings and the implementation of the model in Tinumpuk village, one of the largest sending migrant villages in Indramayu district.

2 Materials and Methods

The study was conducted in Tinumpuk Village—Indramayu District in West Java, Indonesia. This village was selected as nearly 70% of the population working as migrant workers. The selection was also based on discussion with stakeholders in Indramayu District, such as Women's Empowerment and Family Planning Board and the Department of Social Welfare, Labour and Transmigration.

The case studies carried out using a Rapid Assessment Procedure (RAP); A technique of collecting qualitative data for the purpose of obtaining information quickly and useful for decision making towards program intervention. This study mainly used primary data. Secondary data used in consideration of relevance to the phenomenon discussed. Data were collected in various ways: Focus Group Discussion (FGD), in-depth interviews, observation, and documentation.

In-depth interview was conducted to household migrant workers, namely households or families in which one of its members (wife and/or husband) worked or had worked as a TKI. Total Households workers who become informants are twenty households. FGD executed several times with different groups at the district level (stakeholders and partners) and village level (community and religious leaders, migrant households) to obtain more comprehensive data on issues related to the mobility of migrant workers and the impact on their family.

3 Results and Discussion

3.1 The Characteristics of Migrants Worker

There are a number of characteristics of Indonesian migrant workers, which are of particular significance: destination countries and predominance of women. Two destinations are dominant—Malaysia and Saudi Arabia—which together account for over three quarters of all migrant workers. Another important feature is the predominance of women, most of whom are destined to be employed as domestic workers.

The strong economic motivation of international migrants to change the fate or to achieve the desired standard of living for himself and his family bolded their courage to face the problem and risk at work in destination countries. They are also willing to part with the family for a long time. Economic motivation of migrants is indicated by remittances to the family at home to improve the family's economic situation. Wage differentials between receiving and sending areas, as well as migrant's expectations for higher earnings in host countries explained population movements (Todaro 1969).

Ravenstein (1985) declared migrant population groups usually are at productive age. As stated by Lee (1966), migration is selective, and a factor that greatly influences is the life cycle. Many studies have shown that younger age groups are more likely to migrate than the other groups. Results of studies in various countries showed 20–34 age group are more likely to migrate than the other groups (Jansen 1970). There are similar trends in Indramayu district. Agreeing to that, there are many productive-aged migrant workers who leave their toddlers to work abroad. This phenomenon can be understood as that young age is considered more adapt and build compare to the older age group (Nasution 2000).

3.2 The Impact of International Labour Migration on the Family

This chapter focuses on the direct impact of international migration on the families of migrants that are left behind in source countries. More specifically, this chapter focuses on the impact of migration on children, spouses, and parents who are left behind. The types of effects examined included family structure, care of children and the use of remittances.

The study found that since most of such migration is non-permanent and that most involves the separation of husband and wife, it has an impact on family structure and composition outcomes reflecting relationship quality such as divorce. In the village, sometimes it can be found that a returning migrant who has saved sufficient funds to build or purchase a house may indeed find his husband with another. Furthermore, Hugo (2002) revealed that it also has an impact on the headship of households.

Clearly, international labour migration causes various effects not only positive, but also negative impacts. The expected positive impacts are associated with the use of remittances sent to the families and all income earned while working abroad. One of the negative impacts that require a lot of study and solution strategic models at the micro level, namely the impact on the family structure and execution of the functions of the family, among other functions that is parenting function, economic function, education function, and the function of socialization.

The absence of a father or mother to the children, notably the absence of a wife, and a mother to the children for a period times has an effect on the implementation of the role of the family, which in turn may hamper the realization of a quality family, and result in the decline of family resilience. This situation also hampers the establishment of small, happy and prosperous family as campaigned by National Population and Family Planning Board.

3.3 The Impact on the Children

It is now well accepted that international migration of a parent or family member can have both positive and negative effects on the children in the home country. First, there is the possibility that remittances sent from abroad will relax the household budget constraint and result in an increase in child schooling, child health, and a corresponding decrease in child labor. Several studies find evidence supporting this hypothesis (Alcaraz et al. 2012; Yang 2008). Yet, researchers have also recognized that parental migration inherently involves parental absence from the home that pose a negative impact on children which may outweigh the positive effect of remittances.

In migrant households with migrant mothers, children were more often in charge of other family members rather than the father. In Indramayu District, the caregivers were close relatives, with a majority of grandmothers. This finding coincides with the study of women migrating to Saudi Arabia which found that the extended family helped in taking care of the children left behind (Hugo 1995). However, based on the discussion with community leaders in the village, they stated that as consequences of the absence of a mother, children are at risk of suffering the consequences of family disruption, receiving insufficient care and control, and even falling in hands of incompetent caregivers. Bryant (2005) in his review on the literature on the impacts of parents' migration on children also found that the negative impacts of parents' absence jeopardized the condition of the children left-behind, even though there is little evidence on how children cope with their migrant parents' absence, and on how their wellbeing is affected.

In Indramayu socio-cultural construction, child care remains the primary task of women. Therefore, this issue requires a response at the community level. Ideally, there should be a child care institution that can effectively and efficiently substituting the role of a mother in term of education and socialization of children since education and childcare often becomes a problem when submitted to an extended family. Consistent with this, Giannelli and Mangiavacchi (2010) found that parental migration has a negative impact on school attendance for children left behind in Albania. In addition, the study also found that there was a difference between children in households with non-migrant and migrant family members. Children in migrant households have identified more symptoms of social problems—like problems with their peers; and psychologically, children in migrant households stated less happier when compared with children in non-migrant families. This finding was similar to studies conducted by Sukamdi and Haris (2000) on the impact of international migration on families and children of migrant workers, of which found that the impacts resulting in the psychological condition of the children.

3.4 The Use of Remittances

The role of remittances in a nation's economic development has been well documented by several contemporary studies. Remittances can help improve the country's development prospects, maintained macro-economy stability and reduced poverty (ADO 2008). At the micro level, remittance is one-economic resources that are important to families in changing socio-economic status of the migrant workers and their families. Among the indicators of success, it goes along with the increase in assets such as paddy fields, houses, motor cycles, and electronics. Abella (1992) on his study which expressed the same view point that workers' remittances received from the Middle East have positive economic and social effects on households.

International labour migration produces a large inflow of valuable remittances for the migrants' home countries. The result regarding remittances use patterns revealed that the migrant families use the major portion of the remittance on household consumption, house improvement, purchase of land, etcetera; but relatively small portion of the remittances was used on productive investment (Azhar 2007).

4 Integrated Model for Strategic Solution to the Family of International Migrant Workers: Response to the Challenging of the Issues

4.1 Characteristics of Model

Actually many efforts have been done by Indonesia's government to deal with the issues of international Indonesia's migrant workers (TKI) by focussing on the target to potential TKI/ex-TKI (Alumni). However, there is still a limited attention which has given to families of TKI. "Integrated Model for Strategic Solution to Family of

TKI" is a model which tries to response challenging population problems due to the impacts of TKI mobility to foreign countries on their families in the origin area.

This solution model is integrated in its nature; in terms of its target which covers all aspects of family life consist of child, husband, caregiver, and potential TKI/Alumni, as well as integrated implementation through the integration of different institutionalized local groups at village level such as community center for integrated services (*Posyandu*), organization for family welfare education (*PKK*), group for pre-school children development (*BKB*), group for adolescent development (*BKR*), group for aged population development (*BKL*), income generating group for family welfare (*UPPKS*), alumni TKI association at village level (*CBO IBU TIN*/ Community Based Organization—Migrant Workers' Association of *Tinumpuk* Village).

Community empowerment, particularly with the target to the TKI family is a selected approach in the development of an integrated model for strategic solution of TKI family resilience. This approach attempts to put community at village level on a position as actors as well as beneficiaries in the process of solution seeking regarding TKI's problems which are occurring in their life environments. In each step of model development, village community have been involved starting with the step of problems identification and needs assessment, up to seeking for problems solution which is found through a workshop of model formulation at village level.

This model focusses on the strengthening of TKI family resilience which is developed thoughtfully by considering family dimensions as an objective and actors for realizing family quality of TKI, namely self supporting/independence, prosperous, life in harmony, fit, and match to natural and social carrying capacity of life environment (Fig. 1).

This model is also developed by considering original local knowledge. Original local knowledge of the village community becomes a basis for conducting different activities, especially at village level where the model is developed from program and development strategy of model which has been fixed. Therefore, before testing the model, a study on patterns of thought of community, community tradition and region/village potential where the model development takes place was carried out (Fig. 2).

4.2 The Strategy of Model Development

(a) Empowerment and strengthening of community which cares to TKI Family

As a sending area region of workers to foreign countries, certainly many ex-TKI (alumni) can be found in this region. The community of TKIs' alumni have a great potential that can be developed in the program to increase resiliency of TKI family by assumption that they have much knowledge on working in foreign countries and they have good experience as well as bad experience or unexpected events during they work in foreign countries, and also

Fig. 1 Training of trainer for cadre and caregiver

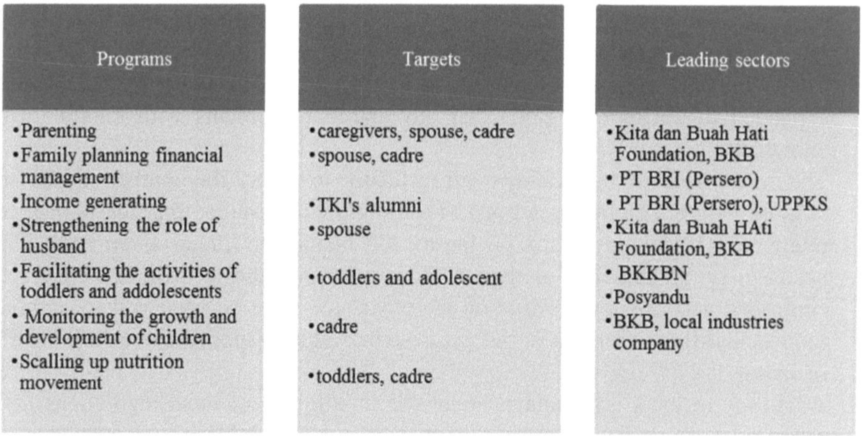

Fig. 2 Design of integrated model for strategic solution to family of TKI

may be shared to other members of community. The TKIs' alumni are expected to be easier to meet TKI family in the village community.

(b) The augmentation of CSR Roles in strengthening of economy empowerment of The TKI Family

In the effort to empower the economy of TKI family, supervision by government and the role of private business in terms of Corporate Social Responsibility (CSR) are conducted to decide enterprise for groups as well as for individual according to willingness and ability of the family, easy to carry out and can be secured its sustainability, so that the selected enterprise can increase income of TKI family. In addition, this strategy is also used to growth awareness of TKI family on their potential in efforts to increase social-economy condition and the use of remittances, and for augmenting ability to utilize the potential of environmental resources efficiently for individual, their families, and also for their environments.

(c) The augmentation of Father's Role and Caregiver

Caregivers have important roles in the growth of child and child development, those are .entrusted to them. Caregivers are expected to understand what, how, different aspects and factors which influence child care, and how to form positive behaviours of children. However, a husband or wife, grandfather or grandmother, uncle or aunt, to whom the child is entrusted by someone who has to be a TKI in foreign countries, certainly he or she is not only confusing in facing this responsibility, but he or she may be did not realize this responsibility.

5 Policy Recommendation

1. Strengthening cooperation and commitment of both partners and stakeholders is indispensable since it is an important factor in determining the success of integrated model for strategic solution to family of TKI
2. The integrated model for solution to family of TKI can be replicated and expanded to the other regions that have similar problems with regard local knowledge.
3. The model needs to be accompanied by efforts to reduce the number of women migrant workers migrating abroad to the minimum level, so that the impact of international labor migration on family left behind no longer as an issue that needs to be addressed. For this, on the one hand, the fertility rate must be controlled at a low level, while on the other side must be created work opportunities in different regions in Indonesia so they able to support a decent standard of living
4. Advocacy to local government about the importance of local regulations that restrict the departure of married women who have children under the age of 2 years to become TKI in view of the importance of the presence of a mother within the first 1000 days of a child's life.

References

Abella, M. (1992). International migration and development. In G. Battisella & A. Paganoni (Eds.), *Phillippine labor migration: Impact and policy*. Quezon City: Scalabrini Migration Center.

Alcaraz, C., Chiquiar, D., & Salcedo, A. (2012). Remiitancess, schooling, and child labour in Mexico. *Journal of Development Economics, 97*(1), 156–165.

Antman, F. M. (2012). *The impact of migration on family left behind*. Bonn: University of Colorado and IZA.

Asian Development Outlook 2008 *Workers in Asia*. http://www.adb.org/Documents/Books/ADO/2008/part020302.asp. Accessed on 9/4/2015

Azhar, I. A. K. (2007). *Overseas migration and its socio-economic impacts on the families left behind in Pakistan, a case study in the Province Punjab, Pakistan*. Kassel: KAssel University Press.

Bryant, J. (2005). *Children of international migrants in Indonesia, Thailand and the Philippines: A review of evidence and policies* (Innocenti Working Paper No. 2005-05). Florence: UNICEF Innocenti Researcher Center.

Giannelli, G. C., & Mangiavacchi, L. (2010). Children's schooling and parental migration: Empirical evidence on the 'left-behind' generation in Albania. *Labour, 24*, 76–92.

Hugo, G. (1995). International labour migration and the family in Indonesia. *Asian and Pasific Migration Journal, 11*(1), 13–46.

Hugo, G. (2002). Effects of international migration on the family in Indonesia. *Asian and Pacific Migration Journal, 11*(1), 13–46.

Jansen, C. (1970). Migration: A social problem. In C. Jansen (Ed.), *Readings in the sociology of migration*. Oxford: Pergamon Press.

Lee, E. S. (1966). A theory of migration. *Demography, 3*(1), 47–57.

Nasution, M. A. (2000). International migration in Southeast Asia: A case study of Indonesian Workers in the Malaysian Peninsula. In Sukamdi & A. Haris (Eds.), *Labor migration in Indonesia: Policies and practices*. Indonesia: Population Studies Center, Gadjah Mada University.

National Agency for the Placement and Protection of Indonesian Migrant Workers. (2012). *The placement of Indonesian migrant workers*. Jakarta: BNP2TKI.

Ravenstein, E. G. (1985). The laws of migration. *Journal of Statistical Society of London, 48*, 167–235.

Sukamdi, & Haris, A. (2000). *Labor migration in Indonesia: Policies and practices*. Indonesia: Population Studies Center, Gadjah Mada University.

Todaro, M. P. (1969). A model of labor migration and urban unemployment in less developed countries. *American Economic Review, 59*(1), 138–148.

UN Women. (2013). *Contribution of migrant domestic workers to sustainable development*. Bangkok.

Yang, D. (2008). International migration, remittances and household investment evidence from Phillippine migrants. *Economic Journal, 118*(528), 591–630.

CSR in the Context of Transition Economy: An Evaluation of Enterprises CSR Practices in China

Bing Zhu and Andre Habisch

1 Introduction

The concept of corporate social responsibility (CSR) has been studied for decades in Western countries (Carroll 1999; Whetten et al. 2002), and highlighted interested in CSR has attracted attention of researchers and scholars worldwide. European Commission (2001, P. 8) defined CSR as "a concept whereby companies integrate social and environmental concerns in their business operations and in their interaction with stakeholders on a voluntary basis". CSR is also defined as "the commitment of business to contribute to sustainable economic development, working with employees, their families and the local communities" (World Business Council for Sustainable Development 2001). Generally, CSR is the management practices by which organization tries to minimize negative impacts of its operation and exert a positive influence on society (Low 2016).

To date, the development of CSR has varied from a pure economic perspective to a perspective of all-sided initiative social responsiveness, which highlights the role of an organization in social system especially for its long-term development (McGee 1998). The concept of CSR has been actualized through philanthropy, minority support programs, cause-related marketing, socially responsible employment and manufacturing, and community organizations' goals advancement (Drumright 1994; Smith 1994). The essential of business today is not only profit maximization but comprehensively sustainable development covering economic, environment and social performance (Rahim et al. 2011). For that reason, CSR is perceived as an effective means to achieve sustainable development. Remarkably,

B. Zhu (✉) • A. Habisch
Ingolstadt School of Management, Catholic University of Eichstätt-Inoglstadt, Ingolstadt, Germany
e-mail: bingzhu.mkt@gmail.com; andre.habisch@ku-eichstaett.de

H. Lu et al. (eds.), *Building New Bridges Between Business and Society*, CSR, Sustainability, Ethics & Governance, DOI 10.1007/978-3-319-63561-3_14

by presenting organization's moral obligation through CSR practices (Pride and Ferrell 2000), business practice will be widely acceptable and recognized by stakeholders like employees, suppliers, community groups, non-governmental organizations and government, etc. The engagement in CSR practices will not only consolidate organizations' legitimacy (Handelman and Arnold 1999) but also help organization gain a positive recognition from both international and domestic community-that is being "social responsible" (Brown and Dacin 1997; Sen and Bhattacharya 2001). Since CSR has been long favored by scholars and business organizations from different context, an understanding of how CSR works differently will help enterprises to be globalized successfully (Amezaga et al. 2013, Alon et al. 2010, Maignan and Ralston 2002, Gjølberg 2009).

A Glance on CSR in China

Since 1978 economic transition in China has been remarkable, and China has been "transformed from a rural agrarian economy to an urban industrial force" (Economic Roundup 2012). The rapid economic growth has resulted in a disjunction between economic development, social development and environment (Hao et al. 2017). A sustainable transition thereby is necessary and important to future development. Sustainable development is an inevitable choice to reconstruct a balanced and healthy system, which can lead to an adjustment of contemporary business strategy among enterprises (Tan 2011, Jefferson 2016). For instance, Chinese enterprises, to a certain extent, believe that CSR-orientated business practice will improve their performance socially and economically rather than focusing on profitability alone (Chen 2009). This change has been manifested through two perspectives.

From business perspective, Chinese firms look for a win-win strategy which will allow them to survive since market competition is getting fiercer than ever in Chinese market. The fact is that Chinese consumers nowadays present their growing concerns on CSR issues through their consumption behaviors. For example, consumers will boycott the firms that produce products with toxic ingredients, overuses chemicals, destroys environment preservation, abuses employees, violate Labor Law, etc. Accordingly, the firms have to make changes to get or to retain their customers by adopting CSR initiatives, perform responsibly for society and firms' products or services. From legalization perspective, CSR was introduced into Chinese Law in 2005, which states that "companies should be honest, keep faith, comply with the social morality and undertake social responsibility" (Chen and Kong 2009, P. 146). Business and economic activities are restricted by SA8000 and ISO14000 (Yu et al. 2010). Based on the findings of an internet survey in 2006, which investigated CSR in mainland China, CSR initiatives with government standard was widely aware and supported by the public and government officials (Ho 2006). For example, the Party secretary of Jangsu Province stated that environment must be strictly protected even though GDP of the province would drop 15% due to the implementation of environmental protection measures (Anderlini and McGregor 2007). Generally, CSR is a useful instrument to achieve sustainable transition and to construct a "harmonious society" in China (Yi 2005; See 2007).

Belal (2001) noted that most of the CSR studies have been conducted from Western cultural context, and there is little about CSR in emerging markets and developing countries like in China. As corporate websites have been a powerful communication instrument to present CSR practices, to create positive corporate image, and to influence public opinions, the researchers aim to measure how CSR practices will be presented through corporate websites in China. In addition, An integrated measurement of CSR practices developed by Maignan and Ralston (2002) is applied in the study. Four research questions will be answered as follows:

Research question 1: To what extent do Chinese enterprises discuss CSR in their websites?
Research question 2: Which motivational principles of CSR are applied most by Chinese enterprises into their CSR practices?
Research question 3: Which managerial processes of CSR are implemented most by Chinese enterprises as their CSR practices?
Research question 4: Which stakeholder issues are mainly concerned by Chinese enterprises in their CSR practices?

Most importantly, the four research questions will be answered in the scope of ownership structure, because ownership structure is an influential factor in CSR practices in China (Lin 2010, Qu 2007, Li and Zhang 2010).

The structure of the paper is as follows. First, the researcher reviews CSR literature. Second, the researcher presents research methodology. Then, the researcher provides the analysis of research findings. Finally, the researcher closes with concluding remarks.

2 Literature Review

In general, CSR has been popular among international scholars. Abreu et al. (2005) revealed the importance of social-cultural factors in CSR practices in Portugal. Broberg (1996) studied CSR practice in Germany, in which strict adherence to the regulation and abiding by the law are the prerequisites for firms to have a good corporate citizenship. A study in American context shows that spirit, law, regulation and social-cultural value are the things that the firms should pay their respect to (Enderle and Tavis 1998). Research by Uhlaner et al. (2004) in Dutch context shows that a combination of CSR perspectives including "economic benefits, conformance to legal and ethical expectation and philanthropic/community involvement" can better explain how CSR orientations vary among family firms.

Specifically, Wood (1991) extended "corporate social performance" model as a framework to analyze corporate social performance, which consists of principle of CSR, process of corporate social responsiveness and outcomes of corporate behavior (John et al. 2006). Maignan and Ralston (2002) studied the difference of how business organizations demonstrate CSR practices through their websites among enterprises in France, the Netherlands, the UK, and the US by applying Wood's CSP model. Based on Wood's model, Maignan and Ralston (2002) developed an instrument to measure CSR practices based on integrated categories including principle of CSR, process of corporate social responsiveness and stakeholder issues. These three elements are explained as follows:

Principle represents the inputs that motivate organizations committing to corporate social responsibility (Wood 1991). It operates at three levels which are institutional, organizational and individual. These components were developed based on social legitimacy (Davis 1973), responsibility for organizational outcomes/impacts (Preston and Post 1975), and managerial discretion exercised by individual "moral actors" from Carroll's discretionary responsibility (Carroll 1979).

Maignan and Ralston (2002) grouped CSR motivation into value-driven, performance-driven and stakeholder-driven based on the suggestion of Swanson (1995). The so-called value-driven motivation presents the concept of CSR running through organization's core value system. Organizations are self-motivated and target-oriented to transition from harm minimization to value creation (Luetkenhorst 2004). Performance-driven motivation, derived from utilitarian perspective, will lead organizations to achieve its financial or marketing performance in terms of profitability, ROI, sales volume, etc. Stakeholder-driven motivation coerces organizations to adapt CSR practices in order to respond to stakeholders' expectation and demands. Main stakeholders including customers, employees, shareholders, suppliers, the government and communities where the firm operates advocate appropriate social and economic behaviors from the organizations. CSR principle has been studied by Weyzig (2006) and Araya (2006) in Mexican context, for example. In their studies, performance-driven motivation is the major force drive Mexican companies engage in CSR practices. Value-driven motivation is common in the U.S. while performance-driven motivation is popular in Europe countries like the U.K., French and Netherlands.

Process is the means through which motivation is transferred to real practices. It presents the process of how the organization identifies and responds to the external environmental factors (threat and opportunity) and the internal management coping mechanism (John et al. 2006). Wood (1991) pointed three types of CSR processes out: environmental management, issues management, and stakeholder management. Study of Maignan and Ralston (2002) extended the content of this step. The CSR process refers to the activities and programs that bring CSR motivation into reality through seven items, which are philanthropy program, sponsorship, volunteerism, code of ethics, quality program, health and safety program and management of environment impact. Logsdon et al. (2006) studied CSR managerial process in Latin American context, in which Mexico engages t in philanthropy program most amongst the Latin American countries. What's more, Maignan and Ralston (2002) studied CSR process from France, the Netherlands, the UK, and the US perspectives, and the companies from UK and US mention code of ethics more as compared to the companies from France and Netherlands.

Stakeholder issues are the issues concerned by stakeholders that are relevant to organizations' activities and operations. Clarkson (1995) divided stakeholders into five groups, which are community, customers, employees, shareholders and suppliers. According to Visser (2007), corporate social investment in health and safety, education and environment is the major stakeholder issue. The study of Isabelle and David revealed that in terms of community stakeholder issues, US firms concern more on quality of life and education as European countries concern more on the protection of environment.

3 Research Methodology

As an influential communication tool, corporate websites indicates plentiful information of CSR initiatives (Esrock and Leichty 2000; Chapple and Moon 2005); therefore, the researchers intend to evaluate CSR initiatives of Chinese firms through corporate websites. According to the statistics from China Internet Network Information Center (2013), by 2012, the internet penetration rate was 39.9%. A 39.9% of internet penetration rate was higher than average internet penetration rate in Asia which was 27.5% (Statista 2012). As a result, the researcher believes that a high internet penetration rate ensures the accessibility of internet users to firms' corporate websites where CSR practices will be demonstrated.

Eighty-five large enterprises based in mainland China on the list of Global 500 (2013) will be involved in this study covering industries of energy industry, iron and mining industry, banking industry, automobiles industry, Telecommunication and IT industry, construction industry, insurance industry, transportation industry, etc. Since these enterprises have great influence on society and economy in mainland China, the researchers believe that they represent the CSR practices from mainstream. The corporate websites that have no English version and only present a specific area of a firm's activity will not be included in this study. Therefore, 80 Chinese enterprises finally will be included in this research and categorized based on their ownerships: central state-owned enterprises (CSOEs), local state-owned enterprises (LSOEs) and private-owned enterprises (POEs).

The economy in China is dominated by state-owned enterprises. After reformation of state-owned enterprises in 2003 (SASAC 2014; David et al. 2006), the state-owned enterprises are divided into central state-owned enterprises (CSOEs) and local state-owned enterprises (LSOEs). 115 CSOEs is directly and wholly under the control and management from the State-owned Asset Supervision and Administration Commission (SASAC), and CSOEs is monopoly in natural resources, medical, aviation, military, petrochemical, metallurgy industries, etc. In 2012, CSOEs contributed 50% of China's goods and service and employed over half of the nation's

Table 1 Measurements of corporate social responsibility practices

Measurement of CSR	Description
Motivational principle of CSR	
1. Value-driven	CSR is indicated as part of company's culture, or as an expression of its core value
2. Performance-driven	CSR is a part of company's economic mission, as an instrument to improve its financial performance and competitive posture
3. Stakeholder-driven	CSR is a response to the pressure and scrutiny of one or more stakeholder groups
Managerial process of CSR	
1. Philanthropic program	Company have a clear mission and application procedures to allocate donations and grants
2. Sponsorship	Company makes it as a type of responsibility imitative aimed at providing assistance either financial or in-kind to a cause or charity
3. Volunteerism	Company has clear program that allows employees to work for a good cause during paid working hours
4. Code of ethics	Company discusses the content and/or implementation of a code of ethics or conduct
5. Quality program	Company has formal product/service quality program
6. Health and safety program	Company sets up a health and safety program to aim at one or more stakeholder group
7. Management of environment impact	Company launches activities to diminish the negative impact of productive activities on the natural environment
Stakeholder issues	
1. Community stakeholder	
a. Art and culture	Company gives supports to organizations, activities, actors and objects linked to the arts or the national culture
b. Education	Company gives supports to the activities to improve the educational opportunities and the quality of the education received by population outside the company
c. Quality of life	Company dedicates to improve the quality of life and well-being of the communities where the company operates, or of society as a whole
d. Safety	Company concerns for the safety of the persons in the communities surrounding its productive operations
e. Perception of the environment	Company concerns for the preservation of the natural environment either in general or in the communities where the company operates
2. Customer stakeholders	
a. Product/service quality	Company shows the achievement of high product/service quality as a part of its commitment to social responsibility
b. Safety	Company concerns for the safety of its customers in relation with its production activities or product/service

(continued)

Table 1 (continued)

Measurement of CSR	Description
3. Employee stakeholders	
a. Equal opportunity	Company shows its commitment to offer equal chance in recruitment and promotion to all the employees regardless of race, gender, age, or handicap
b. Health and safety	Company concerns for protection of the safety of employees in the workplace along with their overall health level
4. Shareholders	Company expresses its commitment to the involvement of stakeholders in corporate governance
5. Suppliers	Company dedicates to give equal chance to suppliers in terms of gender, race, and size and/or to assuring suppliers' safety

labor force. LSOEs are under the control of local government and central ministries, or majority of share belongs to the government. As for POEs, they are established, owned and controlled by individual people or commercial companies. POEs are fast-growing in recent years and contributes more than one-third of GDP. However, POEs still are not extended to the level of encouragement or reward to transition same as the SOEs (David et al. 2006).

The data will be analyzed by using Microsoft Excel and SPSS version 19, and Chi-square will be applied to analyze if there are significant statistical differences in the listed enterprises reporting the indicators of the measurement instrument. Overall, 21 measurement items that have been applied to examine CSR practices in different context (Maignan and Ralston 2002, Amezaga et al. 2013) will be applied in this study. Table 1 presents these 21 measurement items of CSR practices.

4 Research Findings

In this section, research findings of CSR practices are presented through two steps. First step is to analyze CSR practices in general, and second step is to specifically answer research questions in the scope of ownership structure.

4.1 CSR Practices in Large Chinese Enterprises

Eighty enterprises listed in global 500 were included in this study and were categories mainly into 8 industries which are energy industry, iron and mining industry, banking industry, automobiles industry, Telecommunication and IT industry, construction industry, insurance industry, transportation industry. The evaluation of CSR practices

Table 2 CSR practices by industry*

Industry	CSOEs ($n = 45$)	LSOEs ($n = 28$)	POEs ($n = 7$)	Total ($n = 80$)
Energy	40% (18)	25% (7)	–	31% (25)
Iron and steel	11% (5)	7% (2)	14% (1)	10% (8)
Banking	2% (1)	25% (7)	14% (1)	11% (9)
Automobiles	4% (2)	7% (2)	14% (1)	6% (5)
Telecom/IT	11% (5)	4% (1)	14% (1)	9% (7)
Construction	7% (3)	–	–	4% (3)
Insurance	–	11% (3)	14% (1)	5% (4)
Transportation	7% (3)	4% (1)	–	5% (4)
Others	18% (8)	18% (5)	29% (2)	19% (15)

Note: *$X^2 = 23.187$, $df = 16$, $p < 0.05$

of each industry was assessed and presented in Table 2. The results firstly show significant differences within these industries ($X^2 = 23.187$, $df = 16$, $p < 0.05$). Overall, energy industry presents its CSR more than the rest of industries with 31%.

4.2 CSR Practices Presented by Ownership Structure

In this section the researchers aim to evaluate if there are statistical differences regarding the CSR practices presented by enterprises symbolizing different ownership structure which are CSOEs (central state-own enterprises), LSOEs (state-owned enterprises), and POEs based on Chi-square test.

4.2.1 Motivational Principle of CSR

In aforementioned section, motivational principle of CSR consists of three elements which are value-driven, performance-driven and stakeholder-driven. Based on the description of each principle, the researchers screen corporate culture and mission from "About Us", "Corporate Culture "or "Sustainability Development" to analyze and identify the principle for each enterprise. Some examples are shown in Table 3.

Table 4 presents the different results among CSOEs, LSOEs and POEs. The differences were found in two motivational principle which are value-driven and stakeholder driven. COSEs obtained highest percentage in value-driven principle, and LSOEs present the highest percentage in stakeholder-driven principle. However, it is noteworthy that there was no significant difference found in performance-driven principle.

Table 3 Presentation of principle of corporate social responsibility

Motivational principle of CSR	Example
1. Value-driven	Integrity, commitment, innovation and dedication (http://www.sgcc.com.cn)
2. Performance-driven	To serve as a bridge between production and market demand to achieve win-win outcomes (http://www.crmsc.com.cn)
3. Stakeholder-driven	Serve our customers, maximize return to shareholders, give staff full pay to their talents, and contribute to the society (http://www.icbc.com.cn)

Table 4 CSR principle

	CSOEs ($n = 45$)	LSOEs ($n = 28$)	POEs ($n = 7$)	X^2	Sig
Value-driven	62% (28)	39% (11)	14% (1)	7.546	0.023*
Performance-driven	64% (29)	46% (13)	57% (4)	2.293	0.318
Stakeholder-driven	24% (11)	57% (16)	57% (4)	8.869	0.012*

Note: *Significant at the 0.05 level

4.2.2 Managerial Process of CSR

The program managerial process will be considered if the information is presented in detailed of how the firm applied CSR practices in this stage. For example, CHEMCHINA (http://www.chemchina.com.cn) discusses its environmental management in terms of key concepts, ISO1400 environment management information system, technologies R&D, and how the whole system can protect environment and improve business performance at the same time. Table 5 provides some examples from listed enterprises.

Table 6 demonstrates the results from seven elements of managerial process of CSR, and there are six statistical differences found in philanthropy program, volunteerism, code of ethics, quality program, health and safety program, and management of environmental impact. In contrast, there is no significant difference found in sponsorship. In terms of philanthropy program, CSOEs present the highest percentage. LSOEs dominate volunteerism, code of ethics, health and safety program, and management of environmental impact with highest percentage. In addition, it is remarkable that private enterprises present highest percentage same as state-owned enterprises in quality program.

4.2.3 Stakeholder Issues

The issues related to different stakeholder groups were evaluated and the results are presented in Tables 7 and 8. The stakeholder issues are commonly shown in the

Table 5 Presentation of managerial process of CSR

Managerial process of CSR	Examples
1. Philanthropic program	Amer International Group presents a Philanthropic program to allocate donation and grants (http://en.amer.com.cn)
2. Sponsorship	Sponsorship program is a part of Geely's responsibility initiatives, through which Geely sponsors various activities in the field of "culture, arts and nature" worldwide (http://global.geely.com)
3. Volunteerism	SINOCHEM group presents a volunteer program which organize employees to make contribution for non-profit organization during paid working hours (http://www.sinochem.com)
4. Code of ethics	China Guodian discuss the role of value-driven principle in its development (http://www.cgdc.com.cn)
5. Quality program	SINOPEC group published "Quality Management Measures" and "Quality-related Accident Management Regulation", adding relevant regulations and the detailed quality management methods with clearly defined rights and responsibility (http://www.sinopecgroup.com)
6. Health and safety program	PICC presents various health and safety programs for its employees and customers (http://www.picc.com)
7. Management of environment impact	CHEMCHINA implement ISO 1400 environment management system to diminish negative impact of its production on the environment (http://www.chenchina.com.cn)

Table 6 Managerial process of CSR

	CSOEs ($n = 45$)	LSOEs ($n = 28$)	POEs ($n = 7$)	X^2	Sig
Philanthropy program	64% (29)	39% (11)	14% (1)	8.568	0.014*
Sponsorships	73% (33)	86% (24)	71% (5)	1.679	0.432
Volunteerism	42% (19)	75% (21)	57% (4)	7.507	0.023*
Code of ethics	56% (25)	82% (23)	43% (3)	6.729	0.035*
Quality program	56% (25)	86% (24)	86% (6)	8.335	0.015*
Health and safety program	31% (14)	89% (25)	71% (5)	24.438	0.000*
Management of environmental impact	53% (24)	93% (26)	71% (5)	12.576	0.002*

Note: *Significant at the 0.05 level

section of "Sustainability", "CSR", "HR", or "Investor Relationship". Table 7 shows some examples of stakeholder issues from listed enterprises.

Out of 11 indicators, 7 indicators show significant differences which are art and culture, education, protection of the environment, product/service quality, safety, health and safety, and shareholder. In the group of community, LSOEs present the highest percentage in the indicators of art and culture and education. CSOEs dominate in the indicator of protection of environment. In the group of customers,

Table 7 Presentation of stakeholder issues

Stakeholder issues	
1. Community stakeholder	
Art and culture	Geely sponsors symphony orchestra (http://www.geely.com)
Education	BAOSTELL Education Foundation was set up to provide various scholarships to students and teachers in order to improve the quality of education in the country (http://www.baosteel.com)
Quality of life	CNOOC group launches a program to aid to Nima county of Tibet region since 2002, which aims to improve quality of life and well-being of Nima (http://www.cnooc.com.cn)
Safety	China Huaneng presents safety as one responsibility to ensure safety of employees in workplace and the community where it operates from four aspects (http://www.chng.com.cn)
Protection of the environment	China Life sets up an tree plantation program in Engebei Desert, Inner Mongolia and Chenjiayu village to protect and improve ecological environment of the Yellow River region (http://www.chinalife.com.cn)
2. Customer stakeholders	
Product/service quality	Dongfeng Motor Group has gained ISO/TS certificate for quality management, ISO certificate for environment management system, health and safety management to ensure product with top quality and to control the negative environment impact on environment in community where the firm operate (http://www.dfmc.com.cn)
Safety	China Unicom established and improved the information detection, monitoring and disposal mechanism **to clean out junk information and clean up network spaces** (http://www.chinaunicom.com)
3. Employee stakeholders	
Equal opportunity	SINOCHEM group presents non-discriminatory human resource policy that treats every employee equally and justly, regardless their nationalities, colors, ethnics, genders, and religious believes (http://www.sinochem.com)
Health and safety	BAOSTEEL presents health and safety program which protect its employees in the workplace (http://www.baosteel.com)
4. Shareholders	Bank of China presents its commitment to its shareholder in corporate governance and issues a series of regulation to preserve the right of its shareholders (http://www.boc.cn)
5. Suppliers	SINOPEC Group expresses its dedication to and acknowledge significant role of its suppliers worldwide. The suppliers are offered an equal chance regardless the size, region and gender (http://www.sinopect.com)

LSOEs obtained the highest percentage in the indicators of product/service quality and safety followed by POEs. As for the group of employees, LSOEs predominate in the indicator of health and safety. Finally, in the group of shareholders, state-owned enterprises also present the highest percentage.

Table 8 Stakeholder issues

	CSOEs ($n = 45$)	LSOEs ($n = 28$)	POEs ($n = 7$)	X^2	Sig
1. Community					
Art and culture	38% (17)	75% (21)	43% (3)	9.788	0.007*
Education	73% (33)	100% (28)	71% (5)	9.153	0.010*
Quality of life	69% (31)	89% (25)	71% (5)	4.064	0.131
Safety	69% (31)	79% (22)	43% (3)	0.816	0.665
Protection of the environment	80% (36)	71% (20)	71% (5)	7.243	0.027*
2. Customer					
Product/service quality	56% (25)	93% (26)	71% (5)	11.444	0.003*
Safety	42% (19)	79% (22)	71% (5)	9.941	0.007*
3. Employee					
Equal opportunity	76% (34)	89% (25)	71% (5)	2.386	0.303
Health and safety	51% (23)	89% (25)	43% (3)	11.342	0.003*
4. Shareholders	42% (19)	68% (19)	57% (4)	21.433	0.000*
5. Suppliers	40% (18)	32% (19)	43% (3)	0.549	0.760

Note: *Significant at the 0.05 level

5 Discussion and Conclusion

In this section, firstly four research questions will be discussed based on the research findings. Then, the discussion will be provided to reveal the main phenomenon of CSR practices in China.

In response to research question 1: the findings indicated in Table 2 reveal that energy industry, iron and steel industry, and banking industry are the top three industries that emphasized CSR practices most through their websites. This can be explained by government action that effective and strict measures to limit high energy consuming and high polluting industry such as energy industry, iron and steel industry has been implemented, which is an important signal to support sustainable transition. The policy and attitude of Chinese government are the most vital forces for business to operate in a responsible way. For that reason, these industries have to adjust their industrial structure to be low energy-consuming and low polluting. The CEOs of these large enterprises believe that participation of CSR is an effective way to adjust industrial structure and will improve business performance and public image. In addition, Table 2 also shows that CSOEs and LSOEs contribute large proportion in CSR practices. Basically, CSOEs and LSOEs in China have to be responsible for social welfare of their current and past employees, and they also play a major role in the region and/or in the communities where they have been operated.

In response to research question 2: the findings shown in Table 4 report that CSOEs contribute larger proportion in value-driven principle, and LSOEs and POEs dominate in stakeholder-driven principle. In addition, since there is no statistical difference in performance-driven principle, performance-driven principle is common principle applied by the enterprises as achieving financial or marketing performance is common goal for the enterprises. Also, LSOEs significantly presents the highest level in management of environmental impact. In order to diminish the negative impact of productive activities in the local context, some LSOEs have updated their purification and sewage disposal system or recycle system, some have launched paperless program, and some have maximized the use of renewable resources.

In response to research question 3: Table 6 revealed that sponsorship was the only indicator that did not present statistical differences, with over 70% obtained in each business nature. This manifests that most enterprises have been attached to sponsorships. What occurs among Chinese enterprises most is to provide financial assistance to the society and the community such as charity, sponsoring culture event and sport competition. LSOEs accounted for the higher percentage than CSOEs and POEs.

In response to research question 4: the findings shown in Table 8 explicate that the CSR practices presented by enterprises are mostly related to community, customers, employees and shareholders. Overall LSOEs performed better than CSOEs and POEs in stakeholder issues, and all LSOEs mentioned education issue in their corporate websites, which shows their initiatives in developing and improving education. They have launched various activities to help the education in villages, remote area and less developing regions such as in western part of China. The activities include buying books for students and schools, constructing Hope Primary Schools, raising funds to grant poor students, etc. This implies that LSOEs focuses more on establishing and maintaining a relationship with stakeholders as LSOEs operates in local context.

In conclusion, this study revealed that enterprises from different business nature in China present dedication of being corporate social responsible at different levels. CSOEs and LSOEs are still the main force of CSR practice in China. The research findings imply that corporate social responsibility has received recognition and application in various industries more than ever before, and CSR practice becomes a mainstream of sustainable development in China.

References

Abreu, R., David, F., & Crowther, D. (2005). Corporate social responsibility in Portugal: Empirical evidence of corporate behavior. *Corporate Governance, 5*(5), 3–18.

Alon, I., Fetscherin, M., Lattenmann, C., Li, S., & Schneider, A. M. (2010). Usage of public corporate communications of social responsibility within Brazil, Russia, India and China (BRIC). *International Journal of Emerging Markets, 5*(1), 6–22.

Amezaga, T. R. W., He, H. Y., & Cuervo, A. A. V. (2013). Socially responsible corporate communications in Mexico: Evidence of the large companies by their operational levels. *International Journal of Business and Management, 8*(10), 144.

Anderlini, J., & McGregor, R. (2007, July 10). Campaign to clean the Yangtze under way. *Financial Times*.

Araya, M. (2006, Spring). Exploring terra incognita. Non-financial reporting in corporate Latin America. *Journal of Corporate Citizenship, 21*, 25–38.

Belal, A. (2001). A study of corporate social disclosures in Bangladesh. *Managerial Auditing Journal, 16*(5), 274–289.

Broberg, M. P. (1996). Corporate social responsibility in the European communities–the Scandinavian viewpoint. *Journal of Business Ethics, 15*(6), 615–625.

Brown, T., & Dacin, P. A. (1997). The company and the product: Corporate associate and consumer product responses. *Journal of Marketing, 61*, 64–84.

Carroll, A. B. (1979). A three-dimensional conceptual model of corporate social performance. *Academy of Management Journal, 16*(2), 312–322.

Carroll, A. B. (1999). The pyramid of corporate social responsibility: Toward the moral management of organizational stakeholders. *Business Horizons, 34*(4), 39–48.

Chapple, W., & Moon, J. (2005). Corporate social responsibility (CSR) in Asia: A seven country study of CSR website reporting. *Business & Society, 44*(4), 415–441.

Chen, X. D. (2009, April 15–17). *CSR in China: Consciousness and challenges. A study based on Zhejiang province.* In Paper for the Conference "US-China Business Cooperation in the 21st Century: Opportunities and Challenges for Entrepreneurs", Indian University, Indianapolis and Bloomington, Indiana.

Chen, H., & Kong, Y. S. (2009). Chinese consumer perceptions of socially responsible consumption. *Social Responsibility Journal, 5*(2), 144–151. doi: https://doi.org/10.1108/17471110910964441.

China Internet Network Information Center. (2013). Accessed December 15, 2014, from http://www1.cnnic.cn/IDR/ReportDownloads/201310/P020131029430558704972.pdf

Clarkson, M. E. (1995). A stakeholder framework for analyzing and evaluating corporate social performance. *Academy of Management Review, 20*(1), 92–118.

David, A. R., Jane, T. T., Robert, H. T., Xue, L. W., & Carolyn, E. (2006). Today's state-owned enterprises of China: Are they dying dinosaurs or dynamic dynamos. *Strategic Management Journal, 27*(9), 825–843.

Davis, K. (1973). The case for and against business assumption of social responsibilities. *Academy of Management Journal, 16*(2), 312–322.

Drumright, M. E. (1994). Socially responsible organizational buying: Environmental buying as a non-economic buying criterion. *Journal of Marketing, 58*(3), 1–19.

Economic Roundup. (2012). *China in a new period of transition.* Accessed August 20, 2014, from http://www.treasury.gov.au/PublicationsAndMedia/Publications/2012/Economic-Roundup-Issue-4/HTML/article2

Enderle, G., & Tavis, L. (1998). A balanced concept of the firm and the measurement of its long-term planning and performance. *Journal of Business Ethics, 17*, 1121–1144.

Esrock, S., & Leichty, G. (2000). Organization of corporate web pages: Publics and functions. *Public Relations Review, 26*(3), 327–344.

European Commission. (2001). Green paper-promoting a European framework for corporate social responsibility, COM 2001, 366, Bruxelles, European Commission.

Gjølberg, M. (2009). The origin of corporate social responsibility: Global forces or national legacies. *Socio-Economic Review, 7*(4), 605–637.

Global 500. (2013). Accessed July 20, 2014, from http://www.forbes.com

Handelman, J. M., & Arnold, S. J. (1999, July). The role of marketing actions with a social dimension: Appeals to the institutional environment. *Journal of Marketing, 63*, 33–48.

Hao, F. Y., Sheng, L., & Pan, J. G. (2017). *Introduction political economy of macao since 1999.* Singapore: Palgrave.

Ho, B. (2006). Confucian businessmen. *CSR Asia Weekly*, 2(43).

Jefferson, H. G. (2016). *State-owned enterprise in China: Reform, performance, and prospects.* Department of Economics and International Businesss School: Brandeis University. Retrieved from http://www.brandeis.edu/departments/economics/RePEc/brd/doc/Brandeis_WP109.pdf.

John, R. D., Kimberly, K. M., & Shruti, G. (2006). The effects of CEO pay structure on corporate social performance. *Journal of Management, 32*, 329–342.

Li, J. W., & Zhang, R. (2010). Corporate social responsibility, ownership structure, and political interference: Evidence from China. *Journal of Business Ethics, 96*(4), 631–645.

Lin, L. W. (2010). Corporate social responsibility in China: Window dressing or structural change. *Berkeley Journal of International Law, 28*(1), 64. Accessed July 22, 2014, from http://scholarship.law.berkeley.edu/bjil/vol28/iss1/3

Logsdon, J., Thomas, D., & van Buren, H. (2006, Spring). Corporate social responsibility in large Mexican firms. *Journal of Corporate Citizenship, 21*, 51–60.

Low, P. M. (2016). Corporate social responsibility and the evolution of internal corporate social responsibility in 21st century. *Asian Journal of Social Sciences and Management Studies, 3*(1), 56–74.

Luetkenhorst, W. (2004, May/June). Corporate social responsibility and the development agenda: The case for actively involving small and medium enterprises. *Intereconomics, 39*(3): 157–166.

Maignan, I., & Ralston, D. (2002). Corporate social responsibility in Europe and the U.S.: Insights from businesses' self-presentations. *Journal of International Business Studies, 33*(3), 497–514.

McGee, J. (1998). Commentary on 'Corporate strategies and environmental regulations: An organizing framework' by A. M. Rugman and A. Verbeke. *Strategic Management Journal, 19*(4), 377–387.

Preston, L. E., & Post, J. E. (1975). Private management and public policy: The principle of public responsibility. *California Management Review, 23*(3), 56–62.

Pride, W. M., & Ferrell, O. C. (2000). *Marketing: Concepts and strategies.* Boston: Houghton Mifflin.

Qu, R. (2007). Effects of government regulations, market orientation and ownership structure on corporate social responsibility in China: An empirical study. *International Journal of Management, 24*(3), 582.

Rahim, A. R., Jalaludin, W. F., & Tajuddin, K. (2011). The importance of corporate social responsibility on consumer behavior in Malaysia. *Asian Academy of Management Journal, 16*(1), 119–139.

See, G. K. H. (2007). Harmonious society and CSR: Transforming the business environment? *CSR Asia Weekly, 3*(38), 8–10.

Sen, S., & Bhattacharya, C. B. (2001). Does doing good always lead to doing better? Consumer reactions to corporate social responsibility. *Journal of Marketing Research, 38*(2), 225–243.

Smith, C. (1994, May/June). The new corporate philanthropy. *Harvard Business Review*, 72, 105–106.

State-owned Asset Supervision and Administration Commission (SASAC). (2014). Accessed August 20, 2014, from http://en.sasac.gov.cn/

Statista. (2012). *Internet penetration rate in Asia compared to the global penetration rate from 2009 to 2012.* Accessed December 15, 2014, from http://www.statista.com/statistics/265156/internet-penetration-rate-in-asia/

Swanson, D. L. (1995). Addressing a theoretical problem by reorienting the corporate social performance model. *Academy of Management Review, 20*(1), 43–64.

Tan, Z. (2011, May 9–13). Sustainable development and green economy in China. Paper presented at the 9th China-EU Round Table, Xi'an, China. http://www.eesc.europa.eu/resources/docs/sustainable-development-and-green-economy-in-china–tan-zhu.doc.

Uhlaner, L. M., van Goor-Balk, H. J. M., & Masurel, E. (2004). Family business and corporate social responsibility in a sample of Dutch firms. *Journal of Small Business and Enterprise Development, 11*, 186–194.

Visser, W. (2007). Revisiting Carroll's CSR pyramid: An African perspective. In A. Crane & D. Matten (Eds.), *Corporate social responsibility* (pp. 195–212). London: Sage.

Weyzig, F. (2006). Local and global dimensions of corporate social responsibility in Mexico. *Journal of Corporate Citizenship, 24*, 69–81.

Whetten, D. A., Rands, G. P., & Godfrey, P. (2002). What are the responsibilities of business to society? In A. M. Pettigrew, H. Thomas, & R. Whittington (Eds.), *Handbook of strategy and management*. London: Sage.

Wood, J. D. (1991). Corporate social performance revised. *Academy of Management Review, 16*(4), 691–718.

World Business Council for Sustainable Development. (2001). Accessed August 18, 2014, from http://www.wbcsd.org/work-program/business-role/previous-work/corporate-social-responsibility.aspx

Yi, X. (2005). *Improving corporate responsibility competitiveness to promote harmonious society*. Accessed December 10, 2014, from http://www.wtoguide.net/html/2005SinoEuropean/14_36_14_633.html

Yu, L. X., Morrison, J., Yu, L., & Jiang, Q. E. (2010). Standards for sustainable development: Sustainable China trad strategy project. Retrieved from Canada http://www.iisd.org/pdf/2010/china_standards_sd_trade_strategy.pdf.

Women Symbolism in Marketing: Are the Human Rights Legit?

Debarati Das Gupta and Aruna Das Gupta

1 Introduction

Women Empowerment has been a burning agenda for almost half a decade now. In India, especially, with the percentage rise of 1255.3% in rape from 1971 to 2013 (Crime Report 2013, National Crime Records Bureau, India), women's role and safety has been under scrutiny and modification. Indian Jurisdiction, too, has many laws under its leaf meant for the protection and prevention of discrimination based on gender in the case of women. But the real question that lies under all these piles of laws and scrutiny is: Is women empowerment happening, slowly or is it at a standstill?

Mention-able laws under the belt of Indian Constitution are—The Immoral Traffic (Prevention) Act, 1956; The Sexual Harassment of Women at Workplace (Prevention, Prohibition and Redressal) Act, 2013. There is also a particular ministry under the Govt. of India, i.e. Ministry of Women and Child Development.

There have been many debates, arguments and revised or new policies regarding Women, their safety, their image and their development and education. One such explicit Policy under the Ministry of Women and Child Development is the National Policy for the Empowerment of Women (2001).

This policy is entitled to carry out a mission for women against gender biases across different areas of societal aspects: Education, Rights of a Girl Child, Mass Media, Violence against Women, Women in various circumstances, Science and Technology, Environment, Housing and Shelter, Drinking water and Sanitation, Nutrition, Health, etc.

D.D. Gupta (✉) • A.D. Gupta
International Management Institute, New Delhi, India
e-mail: debaratidg@gmail.com; arunadg26@gmail.com

H. Lu et al. (eds.), *Building New Bridges Between Business and Society*, CSR, Sustainability, Ethics & Governance, DOI 10.1007/978-3-319-63561-3_15

This policy described as an integral part of the Principle of Gender equality that in turn is intensely imbibed in the Preamble, Fundamental Rights and Fundamental Duties and Directive Principles.

However, women projection in media is far from following this policy. Freedom of Press is another such right given to the press; this right's misuse allows the press/media to market any symbol of a product showcasing women. Such a misuse of a policy or rule overpowers and violates the National Policy for Empowerment of Women (and in turn, Fundamental Rights). This paper intends to show, briefly, how marketing strategies have been violating the Fundamental Rights (and different Policies). These rights are imbibed into the Constitution of India and acknowledged to be a Right for every woman (and also, every human).

A keener view into the details of the Fundamental Rights of India, guarantees two things: Firstly, that it is enforceable to all human beings (of India) and secondly, if in any violation of these rights, that individual or group of individuals can take it up with any court of law in India. The Fundamental Rights guarantees 'Right to Equality' i.e. there shall be no discrimination in terms of caste, creed, sex or place of birth. Similarly everyone is guaranteed an equal view from the perspective of law. Third is the Right of all citizens in terms of Public Employment. Now our focus is mainly on the Right to Equality—and in what conditions have they been violated via marketing strategies. The media publishes without doubt via the Freedom of Press Act whatever the advertisements concur. These advertisements encroach upon the symbol of gender equality and question the maintenance of these Fundamental Rights, especially the Right to Equality.

Every organisation or company that functions in India must adhere to the country's policies and take care that it doesn't encroach upon any rules or policies. However, the Freedom of Press Act gives such companies freedom to misuse this right and market ideas that distinguish directly or indirectly between the male and female gender. In a situation of ever-growing rape cases and violence against women, a lot of sectors (media, society, corporate sector) have to conceive together a notion of responsibility. This also comes under the belt of Corporate Social Responsibility as it affects in the procurement, settlement and acceptance of the image they sell in advertisements.

2 Cases Studied

We take up each case from the viewpoint of gender bias in terms of demeaning one gender over the other or using to market a product via projection of a particular sex as a commodity.

Case 1: Amul Macho Innerwear Ad Aired on Television (2007–2008)

What the Ad carried: This ad garnered a lot of controversy due to its sultry atmosphere. A woman is shown to wash her husband's clothing including his innerwear, and while doing so, makes sexually arousing expressions. Onlookers (other women shown here) look at her and express a wide-eyed awe and wonder with an expression that features the word 'toing'. Now the word 'toing' was to suggest the durability of the innerwear. However, due to the ad's ambiguity and sexual portrayal of the product, it caused the masses to relate the word 'toing' to the man's (here, the woman's husband in the ad) masculinity. On viewing the Ad, it is clear that the marketing strategy moves away from portraying the durability of the product and steps towards the portrayal of a married couple's sexual innuendo.

How the Company reacted: The ad had its run for quite some time and was viewed by a majority of the population only to be banned altogether approximately by mid-2007. Most people reacted to it very strongly saying it is confusing as to what the Ad represents—the product or the image of the couple's privacy. However, it isn't about as much of the content as much as it is in the way of portraying it. This Ad clearly moves away from its agenda of selling the product it adheres to.

Case 2: J. Walter Thompson India leaked Ford Ad Controversy (2013)
What the Ad carried: The uproar with this Ad came quickly when the Ad got uploaded onto the website 'Ads of the world'. The Ad featured three women wearing skimpy clothing, tied and gagged in the trunk of Ford Figo hatchback.

How the Company reacted: The Ad was not accepted by Ford but was rather created by JWT India. JWT India handles the advertisements for Ford in the country. According to the spokesperson of Ford, such an Ad was created without any client approval. This Ad, which would called a 'Fake Ad' as it wasn't distributed on Client's approval. However, the Ad garnered a lot of controversy due to its harsh context on the image of women and how to treat them.

Case 3: Fastrack Sales Ad (2013–2014)
What the Ad carried: The ad carried the words '20% Off Watches, Bags and Sunglasses' alongside a young woman nude, with her body parts covered with yellow sticker rolls on which 'Sale' is written. The visual image portrays the woman as an object to be sold as she is marked with the words 'Sale' over her. This is another example of diversion from the propaganda and proposition of the Ad, which intends to sell watches and bags. The question then arises, how 20% Off bags, watches and sunglasses is equivalent to a woman covered with yellow tape with the words 'Sale' all over her.

How the Company reacted: Although Fastrack's agenda over its past few ads has been 'Move on' which urges the youth to come out of the closet, and not stick to things that are limiting. This Ad clearly misses the mark. The ad garnered a lot of flak and was eventually banned due to protests by NGOs and public in general. Fastrack took the Ad off soon.

Case 4: Moods Condom Ad (2015)
What the Ad carried: This Ad was posted during ICCI Cricket World Cup 2015 on the Facebook product page of Moods Condoms on February 15th. Now condom ads of course market a product that is about intimacy. But sometimes such ads crossover and means something else, and defer from their real goal. Like in this Ad, the company wanted to congratulate Team India for its win against another Cricket

team. In doing so, the ad portrayed a frail young woman in her intimates lying on a coach with a sexually aroused expression and the title of the image as 'Nailed it!' and with the tagline—'That's how to #PlayitRight; Congratulations Team India'. Now this makes the Win of a Cricket team or rather a Sports team to be equivalent to a woman in a dormant position intimately. This portrayal raises a question then, how do these marketing strategies equate these two unparallel areas onto one podium. How does a woman in a sexual position relates to the loss of a winning team is definitely a topic to ponder upon while the rest of the country battles for gender equality.

How the Company reacted: There has been no reaction from the company as such.

3 How These Ads Encroach upon the Fundamental Human Rights

Since many Ads were distributed via the internet, it violated the **Information Technology Act** (2000) which states under Chapter XI, that any material that is lascivious or appeals to the prurient interest may be imprisoned up to a term of 5 years or more. Now this Act, indeed a form of expression is protected under the Article 19(1) a. Transmission of this sort that causes annoyance or 'grossly offensive' are also included under Section 66A. Deemed as an expression, of course, makes it come under the Freedom of Expression right which has stood misused in many ways before. But ads of this sort violate the Gender Equality clause in the Fundamental Rights. But it becomes easy for companies or media to transmit such ads very easily via the Freedom of Press Act. However Freedom of Press Act, under Indian Law, does not confer an absolute right to express one's thoughts freely. Clause (2) of Article 19 of the Indian Constitution enables the Legislature to impose restrictions on free speech under specified categories, of which, two of the categories are 'Defamation' and 'Decency and Morality'.

Now how this comes under Human Rights is a good question. Gender Equality, a sensitive issue in India, for which many NGOs, organisations and the Government, in general are fighting to try and achieve in the society, is a Fundamental Right under the Indian Constitution in its Preamble. These Fundamental Rights apply to all the citizens of India, who, obviously are human beings. Hence it would not be wrong to deem these Fundamental Rights as Fundamental Human Rights of the citizens of India. When such Ads show up, be it through the internet, or through the media (print/news/etc.), it does in fact violate the basic Fundamental Human Right to equality of gender, as one of the genders, here 'Female' is shown under a very disrespected light. Judging each of the Ads, taking the first case of JWT India's Ad, gagged and tied up women give the impression that sexual violence is dubbed correct, as women appear to be of a 'delicate' gender. This does nothing to contribute towards Gender Equality, which as stressed before, is a Human Right, for all humans,

Another question is usually Social Media platforms ban anything obscene from being posted if it feels it manages to divert people's attention in a wrong way. This clearly did not happen in this case although this Ad does violate the Information Technology Act (2000) where a lascivious image is portrayed deferring from its goal and moving towards sexually arousing its viewers. Also, it does nothing to speak about the product of the company, i.e. condoms. This ad amongst all other condom ads clearly diverts the basic purpose of the product, which is 'protection while consensual intimacy'. But the Ad rather focuses on how a win over a team is equal to image portrayed by a woman on a couch in a desirable intimate position.

Such Ads not only move further away from their primary purpose of marketing the product, but so casually uses the female gender to portray unrelated taglines and messages. Similarly, the Amul Macho Ad is an advertisement of the garment. The garment, however, becomes lost in the Ad, as people are left focussing on the woman and her provocative expressions. Hence again, a diversion is caused for the viewer. It does help to arise and point out a strong fact: that marketing agencies and their clients feel that in order to sell a product, a diversion is required to be set up, as in the case of the above-stated examples and many more such as Wildstone Perfumes, Axe Perfumes, Shaving Cream Advertisements, all of them seem to require a distraction—and that mode of distraction is the female gender.

Female Gender seems very easy to exploit as exploitation, in the case of India, and many countries, starts from the embryo itself, with a large ratio of female foeticide cases, following domestic violence, assault and rape. It is not going to be easy to push the wheel of Women Empowerment if people don't start feeling something wrong about the images in the media.

And the most valuable resource of spreading the equality of gender message is by removing and editing of images that pertain to these criteria:

1. Diversion from product value and vision.
2. Unnecessary attribution or usage of a provocative innuendo unrelated to the product.
3. Strict hold over allowing such advertisements to be transmitted via any media platform via Policies, Acts that are already instated in the Constitution of India.

A market, be it over the television set up, or over the internet—is a place where every human being comes onto the same platform. Media is accessible to everyone, from village to urban metropolitan cities. Hence apart from focus groups targeting only crime against women or female foeticide, one must realise how this image got created and transmitted mentally across a 1.2 billion population.

Images play a very covert role in our lives as we are surrounded by them all the time, and the brain processes these images as presented by the media or web, subconsciously. Many people may argue as to how this is related to Fundamental Human Rights. Of course, they are as they violate all the Fundamental Rights of a human being of being upheld equal to another human being, without bias. Hence, the issue of Human Rights is legit, and especially so, in such Ads. Such images violate one particular gender and put it in a variable position. Hence, marketing strategies must base on ethics, value and projection of the product.

4 Proposed Solution

Although much imposition on Freedom of Speech isn't required, certain measures may be upheld by both parties, the Government in question and the Companies (Marketing agencies and Client). Since Human Rights, especially from the perspective of Gender Equality, are in question as to whether they are upheld or not, the Government can easily mandate owing to its many interlinked policies as we've stated above, i.e. Legislation due to lapses in decency and defamation clause over Freedom of Press, Information Technology Act, which are both forms of Freedom of Expression, which comprises an important part of Fundamental Human Rights, of which, Gender Equality is an integral part. While the Companies can improvise on their marketing strategies also taking the Corporate Social Responsibility in perspective. The CSR of any company isn't just through ethics in product manufacturing or supply, CSR is also a part of the company's responsibility towards society and how it helps to build a civic, ethical, integrated society.

Many companies have already shown the way in an effort to bring about ethical transformation of image of gender bias, such as Dove, in its 'beauty campaign' wherein it has shown various ads featuring and focusing on making the skin better and does not talk about fairness and bleaching creams like Fair and Lovely, which although, isn't banned, continues to portray a dismal picture of apartheid society which still believes in discrimination of women and men on the basis of their skin colour.

Dove also, via a popular video sharing platform has brought about considerable focus on the actual truth of beauty where the video showcases how makeup and photo-editing resources are employed to digitize and create a fake picture of female beauty. The campaign emphasizes on believing in real quality and beauty rather than the many digitally enhanced images advertised around the globe. Moving closer to India, Nirma, another company has always maintained its reputation in showcasing ethical ads emphasizing on women strength and equality.

One such ad of Nirma shows how four women don't worry about their clothing getting dirty due to mud while they help push out a stuck ambulance from the mud. The crowd watches on, as they accomplish in pulling the van out and walk away confidently. These sort of advertisements focuses on two things: first the product here, Nirma Washing powder which is capable of washing away all dirt and allows such independent and confident women to not worry about dirt on their clothes, while helping someone in need. In doing so, they also project a positive light on the female gender thus maintaining ethics and integrity.

Similar to Nirma is also another Indian enterprise, Bharti Airtel, which showcases an Ad where a woman is a boss, and her husband is the employee. This ad moves on to build that while the husband is stuck up with some work, his loving wife is just his wife at home and not his boss. She cooks his favourite dishes and waits for him to come back while video chatting with him and showing him the dishes. This Ad does two things too, one it shows that a career position flip is very much possible and there shouldn't be a problem as gender equality is a right to be maintained. Whether it is the woman who is the boss or the man, it shouldn't matter. Also, it showcases the emphatic relationship people have with internet as they use it on a daily basis to reach out to the people they truly cherish. Titan, another company, has ethical ads as well emphasizing on equality in life and work for women and men both while showcasing its flagship product, watches. Such is the power of advertising and marketing, which is undermined. But if used, ethically and resourcefully, can change and grow a lot—but of course with conjugation of the receiver and the giver.

Hence, the solution isn't based on an easy path, both parties in question must work hand in hand via the Public Policy-Private Sector partnership wherein maintenance of Fundamental Rights, especially Human Rights, in terms of no gender exploitation or bias must be employed.

A mutual partnership between the Public Policy and Private sector shall strengthen India's fight for Women's Rights and Gender Equality. The country will be able to uphold its commitment to such issues as it is also an important member of the United Nations Human Rights Council.

References

Crime Report. (1953–2013). http://ncrb.gov.in/CD-CII2013/snapshots%2053-13.pdf

Freedom of Expression in India. Wikipedia.org

Gazette of India – Part II Section I, Ministry of Law, Justice and Company Affairs (Legislative Department) – Information Technology Act (2000). http://www.dot.gov.in/act-rules/information-technology-act-2000

Law Commission of India, Hundred-First Report on Freedom of Speech and Expression under Article 19 of the Constitution. http://lawcommissionofindia.nic.in/101-169/report101.pdf

ORF Issue Brief: Media Freedom and Article 19, Issue Brief # 53, April 2013.

Part III Fundamental Rights Of India. http://lawmin.nic.in/olwing/coi/coi-english/Const.Pock%202Pg.Rom8Fsss(6).pdf

Index

H. Lu et al. (eds.), *Building New Bridges Between Business and Society*, CSR, Sustainability, Ethics & Governance, DOI 10.1007/978-3-319-63561-3

MIX
Papier aus verantwortungsvollen Quellen
Paper from responsible sources
FSC® C105338

If you have any concerns about our products,
you can contact us on
ProductSafety@springernature.com

In case Publisher is established outside the EU,
the EU authorized representative is:
Springer Nature Customer Service Center GmbH
Europaplatz 3, 69115 Heidelberg, Germany

Printed by Libri Plureos GmbH
in Hamburg, Germany